Metamorphosis and Place

Metamorphosis and Place

Edited by

Joshua Parker, Lucie Tunkrova
and Mohamed Bakari

CAMBRIDGE
SCHOLARS
PUBLISHING

Metamorphosis and Place,
Edited by Joshua Parker, Lucie Tunkrova and Mohamed Bakari

This book first published 2009

Cambridge Scholars Publishing

12 Back Chapman Street, Newcastle upon Tyne, NE6 2XX, UK

British Library Cataloguing in Publication Data
A catalogue record for this book is available from the British Library

ISBN (10): 1-4438-1104-1, ISBN (13): 978-1-4438-1104-0

TABLE OF CONTENTS

INTRODUCTION

JOSHUA PARKER

If this volume's theme initially strikes readers as vague, its title's two terms should not. "Metamorphosis" derives from *metá*: "with," denoting "sharing, joint action, pursuit, quest, (and especially) change," and *morphos*: "shape" (Hoad). Juxtaposed with "place," it suggests that different places share characteristics in terms of shape. It also suggests changes in one's own perceptions of a place, or even that, in certain places, we ourselves are changed along with our shifting viewpoints. Place is, after all, simply space charged with human value (Tuan), and if we, as humans and scholars, tend to carry our own values with us, projecting them onto the spaces in which we find ourselves, we also gather and absorb values from the places we inhabit. We change places and are changed by them. This volume hopes to shape the joint action of describing some of those changes.

With critical calls for a post-national or transnational study of culture and literature (Kristeva), one might argue that the only barriers to academics' focusing more systematically on setting, or, indeed, to establishing schools for the study of novels set in mountains, films set in prairies, or poetry dedicated to deserts, are our own language barriers. Claims that we are moving into an era of post-nationalism are perhaps premature. Yet in a sense, we live in a "borderless" world, if for no other reason than that we carry mapping skills from our original cultural geographies with us as we travel. "Perhaps every textual construction of place implies [...] a mapping or symbolic re-presentation of an interior terrain," writes J. Gerald Kennedy (6). Place, like metaphor, is a means of coming to terms not only with the Other, but with the projected changing, questing, pursuing forms of the self. Personal identity, like national, regional, cultural, or disciplinary identity, superimposes its own ideas of geography onto whatever land an author, poet, photographer, songwriter or video game player is dropped into. One travels, wrote Claude Lévi-Strauss, to find away from home something whose presence at home is no longer recognizable. One's own identity is thus sometimes more clearly mapped abroad. The very outlines and contours of one's own psyche are drawn in one's descriptions of an Other which one temporarily inhabits.

Place, Michel de Certeau writes, involves the order in which "elements are distributed in relationships of coexistence. [...] It implies an indication of stability" (117). Writing on the metamorphoses of a concept whose hallmark is stability was the task set for this volume's authors, who have graciously agreed to coexist in its pages. This collection brings together an eloquent group of scholars from a broad range of disciplines (literary and cultural studies, history, political science, architecture, anthropology, art and art history, communication, sociology, medieval studies, lexicography, comparative literature, linguistics, psychiatry and theoretical psychoanalysis), each situated differently with respect to the questions raised by our title's two terms. Their essays proceed from papers presented at the second annual conference on literary and cultural studies, hosted by the English Language and Literature Department of Fatih University in Istanbul, a city that has already undergone numerous metamorphoses in these two years since the conference's presenters met in May 2007.

The conference's interdisciplinary scope explored a wide spectrum of questions. While personal, regional, and national identities may be constructed in terms of place, what happens to a place as personal or societal values and identities change? How do our values change as places themselves are transformed? These questions can be asked of almost any discipline, whether one is taking a photograph, mapping a literary topography, tracing linguistic change in a geographic region or outlining language's importance to our conception of a political territory, building a place of worship on a physical plot of land, constructing it from words on a page, or on a screen with computer software.

Few places are ever uniquely our own. We share them. To project one's own cultural (or disciplinary) markers onto a landscape which is already hardly blank, but full of others' meanings—to know that geographic points stabilizing one's own identity serve, on their reverse side, to support an entirely different set of meanings for others—is an act de Certeau calls a tactic. While attempting to combine the perspectives of various and diverse disciplines is perhaps symptomatic of a certain fantasy of a utopic integration of disciplinary approaches, each of the articles here manages, quite thoughtfully and in its own right, to shed light on how place is both transforming and transformed. Area studies, proposes Hans Kuijper, is only truly interdisciplinary when communication between participants in cooperation from different disciplines allows them to "change their ways of thinking and develop a new one (by politely challenging, and actively listening to, each other)" (215). This is the metamorphosis for which this volume hopes to provide a platform.

Works Cited

Certeau, Michel de. *The Practice of Everyday Life.* Trans. Steven Rendall. Los Angeles: University of California Press, 1984.

Hoad, T.F. *The Concise Oxford Dictionary of English Etymology*. Oxford: Oxford University Press, 1986.

Kennedy, J. Gerald. *Imagining Paris: Exile, Writing, and American Identity*. New Haven: Yale University Press, 1993.

Kuijper, Hans. "Area Studies versus Disciplines: Towards an Interdisciplinary, Systemic Country Approach." *The International Journal of Interdisciplinary Social Sciences*, 3.7 (2008) 205-16.

Kristeva, Julia. *Nations without nationalism*. Trans. L. S. Roudiez. NY: Columbia University Press, 1993.

Lévi-Strauss, Claude. *Tristes tropiques*. Paris: Plon, 1958.

Stavrides, Stavros. "Navigating the Metropolitan Space: Walking as a form of negotiation with otherness." *The Journal of Psychogeography and Urban Research*, 1.1, September 2001.

Tuan, Yi-Fu. *Space and Place: The Perspective of Experience*. Minneapolis: University of Minnesota Press, 1977.

THE LEGACY OF ARTHUR RAPER

LOUIS MAZZARI

Arthur Raper, born at the turn of the twentieth century, was a white, southern farmboy who helped begin the Civil Rights Movement through his study of the roots of lynching during the Great Depression and his research into the South's complex and debilitating race relations. Much of his work was accomplished under the auspices of Franklin Roosevelt's New Deal and the Atlanta-based Commission on Interracial Cooperation.

Raper wrote a great deal about the South, and his writing—from books called trailblazing by the *New York Times* to his articles published in small-town newspapers—was read by common white folk as well as northern black leaders, by both social scientists and labor activists, and his research was cited in *Life* magazine and in the Supreme Court's decision on Brown v. Board of Education, the landmark case that ended the legal segregation of American schools. His work inspired many white southerners to speak out against racial bigotry—and it definitely infuriated more than a few of their fellows.

Raper was certainly no novelist or craftsman of fiction. But as a good documentarian, he became an accomplished storyteller with an eye for some detail or offhand comment that opened onto the dynamics of a social scene or situation. Along with his books, he now-and-then published detailed observations or parables in magazines and journals that illuminated the way racial etiquette entwined with economics in small southern towns. The best of these observations is "Buy White," published while he was researching *Sharecroppers All*, a critique of the South's racial and social caste system. "Buy White" furrows a seldom-plowed field by combining the imagination and detail of short fiction with the eye for social patterns that marks a good sociological study. Over and over, throughout his life, Raper was drawn to trace out the subtleties of Jim Crow in his books and in his day-to-day exchanges, at his typewriter, and in the barber chair.

"Buy White" by Arthur Raper

About three years ago, I went into a shop to get my monthly hair cut. As the clipped hair fell loosely on the white cloth, we talked about all sorts of things—politics, the weather, the movement of country dwellers to town, and the growing soup lines. Mr. Brown, a close friend of mine, chanced to pass by the window.

In studied casualness, the barber inquired, 'Who cuts Mr. Brown's hair?'

I knew that a Negro barber at Herndon's on Peachtree Street did his work, and I merely commented, 'Well, somebody does better by him than I give you the change to do by me. Why, he gets a haircut every week or ten days.'

'Well, he must have a good barber.'

'Yes, his head always looks nice.'

'I've noticed how well his head looks all the time, and I said to myself, 'Now that man has a good barber, and he stays with him.'

'He tells me that he has been using the same barber since he came to Atlanta ten years ago.'

'Well, I surely don't blame him.'

'I think,' finally answering the question asked me, 'that Mr. Brown has his barber work done in Herndon's shop.'

'Some years ago,' continued the barber, 'I had a customer who came to me for shaves, massages, and shampoos. I shaved him every day, and massaged and shampooed him every week, but I never cut his hair. Finally one day I said, 'Mr. Smith, I don't want to pry into your business at all, but I notice that you always have a good haircut.'

'Mr. Smith said, 'Yes, I have been going to a Negro barber at Herndon's for twenty-three years.'

'And I said: 'Well, you always have a good haircut.'

Then, one Saturday afternoon when Mr. Smith came in and was sitting down in the chair I said, 'Well, you have had another one of these good haircuts. It surely is a good one.' And I went on about massaging him, and told him about a friend of mine who liked steaks. I told him there were places all over town where you could buy and eat steaks, good steaks—places on Peachtree run by white people, and that people who bought steaks there helped these white people maintain a decent standard of living.'. . .

By this time the barber was dusting powder on my neck, and as I got out of the chair he chuckled and said, 'You know, I've been cutting Mr.

Smith's hair ever since. That was three years ago; he never had realized before that a white man ought to trade with white people.'

As I left the shop I realized the barber must have told me the story in the hope that I, too, would understand that white people ought to trade with white people and that I would influence my friend Mr. Brown to come to him or some other white barber.

A few decades ago practically all the barbers in Atlanta were Negroes. From year to year they have lost their trade to white barbers. The same thing has occurred in many other fields, and has recently extended to that of such menial tasks as bellhops in Atlanta hotels and curb waiters at the leading roadside stands. The Negro has been shorn of many of his erstwhile economic opportunities, in some instances by organized threats, in others by the shrinkage of jobs, and in still others by methods similar to that utilized by the white barber who used to cut my hair every month.

Work Cited

Arthur Raper. "Buy 'White'." *Far Horizons* (May 1936): 197. Southern Historical Collection, Wilson Library, The University of North Carolina at Chapel Hill (SHC), folder 47.

"TRANSMUTED BY TIME'S HANDLING": METAMORPHOSIS IN JAMES BRANCH CABELL'S *JURGEN*

ELŻBIETA FOELLER-PITUCH

Abstract

In *Jurgen* (1919) the once famous and now neglected Virginian writer James Branch Cabell has created a modernist fantasy of satiric metamorphoses to ironically expose the human longing for order and meaning, revealed in the ideals of love, heroism, and religion. The novel eloquently expresses post-World War I disillusionment, but in a comic and parodic vein which looks forward to the postmodernist fabulists of the 1960s and 1970s. In Cabell's hands the novel morphs into an elaborate medley of classical myths, Russian folklore, Arthurian romance, and medieval dream vision. The traditional folktale form, however, is undermined by Cabell's parody of Frazerian and Freudian theories, as well as his twentieth-century attitudes of sardonic disenchantment. The protagonist, a hen-pecked pawnbroker and erstwhile poet, Jurgen, is a reluctant Odysseus and Faust figure, a trickster hero who receives the gift of youth, while retaining the consciousness of a 40-year-old. His metamorphosis from middle-aged bourgeois to youthful knight is marked by a year-long quest that takes him out of his medieval European homeland and into mythical realms. Transforming himself into a Duke, then Prince, King, Emperor, and Pope, Jurgen moves through a variety of literary spaces— medieval romance, Arthurian legend, Frazerian fertility lore, classical myth, and finally versions of the Christian hell and heaven. Cabell's narrative subverts the traditional role of the hero, as the rejuvenated Jurgen, a jaunty figure and self-proclaimed "monstrous clever fellow," finds he cannot sustain the demands and possibilities of his youthful body in juxtaposition with his world-weary, aging mind. *Jurgen* is a commentary on art and life from the perspective of middle age. It is a *Bildungsroman à rebours*, an education in how to accept maturity, with its waning of possibilities and loss of illusions. The novel deploys a remarkable range of metamorphoses of time and place so as to reveal our tendency towards superimposing meaningful patterns on existence; in this way *Jurgen* allows us to both laugh at our capacity for self-delusion and celebrate our creative urges. Metamorphosis is shown to be the primary principle of life—we are all "transmuted by time's handling" and so are the spaces we inhabit, both outside us and, most importantly, within our minds.

The novel *Jurgen: A Comedy of Justice* (1919) enjoyed *a succès de scandale* in 1920 when it was suppressed (in its third edition) for alleged obscenity. In 1922, after a lengthy and much-publicized trial, a New York judge decided that the novel was not "lewd, lascivious, indecent, obscene and disgusting," and that indeed it had "unusual literary merit" (quoted in Rubin 278). As a result James Branch Cabell, a Virginian writer based in Richmond and a good friend of Ellen Glasgow, became a popular author with an intellectual following during the Twenties, only to fade back into obscurity in the Thirties, when the Great Depression changed the mood of the country. He was still valued by a small, faithful band of afficionados, including the science fiction writer Robert Heinlein (Patterson). A literary periodical devoted to Cabell's staggering oeuvre of over 50 published works came out somewhat irregularly between 1968 and 1996 under the title *Kalki: Studies in James Branch Cabell.* However, to a larger literary public as well as to most literary scholars Cabell is practically terra incognita, though he merits reading for his urbane and witty writing.

Jurgen addresses in fantasy form, as Hemingway's and Fitzgerald's works did in a more realistic mode, the world view of the years following World War I and its attendant disillusionment, loss of stable values, and longing for "a good time." A cautionary tale about the illusions of youth and the disappointments of middle age, it presents us with a metamorphosis of the self from age to youth and back again, revealing throughout "the vivid sense of Time as the great ironist and destroyer" (Davis 90). His hero is both a reluctant Odysseus and a hesitant Faust figure. Cabell's settings are places of the mind, literary locations, mental constructs. And so Cabell excludes any mention of the American South, but presents a thinly veiled satire of the United States as the conquering kingdom of Philistia that spreads mediocre conformity, utilitarian attitudes, and anti-literary bias. The author counters these Philisitine tendencies by combining his formidable erudition with a wryly ironic look at the universal human condition of aging. He constructs an elaborate parody of a medieval quest romance that comments on human nature and its dependence on illusions. As Desmond Tarrant sums it up: "Cabell uses everything that the imagination of previous ages has created, from ancient classical mythologies to the chivalric romances of the France of Chrétien de Troyes and the England of *Gawain and the Green Knight*; Arthurian legend and the universe of Milton meet with the fairy land of Spencer and the voluptuous nights of the Orient" (135). The traditional form, however, is belied by the twentieth-century attitudes underlying it—disillusionment and a wry tolerance for all beliefs as equally invalid, but necessary to man's psychic comfort. In *Jurgen* Cabell has created a modernist fantasy of satiric metamorphoses of time, place, and person to ironically expose the human longing for order and meaning, revealed in the ideals of love, heroism, and religion. Not unlike T.S. Eliot, Ezra Pound, and even more so James Joyce, Cabell looks to the past and uses fragments of the past to create literary work for the present. His novel eloquently expresses post-World War I disillusionment, but in a comic and parodic vein which looks forward to the postmodernist fabulists of the 1960s and 1970s.

In Cabell's hands the novel morphs into an elaborate medley of classical myths, Russian folklore, Arthurian romance, and medieval dream vision, ironically flavored by parodic treatment of Frazerian and Freudian theories, as well as twentieth-century attitudes of urbanely ironic disenchantment. This is a fantasy world several times removed, for Cabell introduces his story as a traditional tale told in an imaginary land. "It is a tale which they narrate in Poictesme, saying: In the old days lived a pawnbroker named Jurgen …" (3). And in this tale the protagonist finds himself undergoing adventures in fantastic locations contrasted to the normalcy of his imaginary medieval European homeland. Jurgen undergoes a metamorphosis from middle-aged bourgeois to youthful knight who finds himself on a year-long quest that takes him through a series of transformed spaces and times. At the end his travels are placed under erasure ("sous rature," to use Derrida's phrase) by being "unimagined" by Koshchei the Deathless, Creator of this world: "For now I have never begun, and now there is no word of truth in anything which you remember of the year just past. Now none of these things has ever happened" (319). However, these "non-events" form the bulk of the novel, thus bringing up questions about the nature of memory and illusion, as well as art and its fictionality.

Jurgen is a middle-aged, henpecked pawnbroker and erstwhile poet who speaks courteously of and then to the Devil and in reward finds that his termagent wife, Dame Lisa, has disappeared. The novel chronicles his reluctant Odyssey to find her. In the process he receives the ambiguously symbolic shirt of Nessus and the gift of youth, while retaining the experience of a forty-year-old; and so he finds himself in the enviable position of uniting both sides of the French proverb: "Si jeunesse savait, si vieillesse pouvait," a saying no doubt Cabell was quite familiar with, since he taught French at the College of William and Mary as an undergraduate there. This combination of old and young Jurgen moves through a fantasy world (or Freudian dream world) in the guise of a conquering hero, involved mainly in sexual exploits, related in amusing *double entendre*. During the solstices he advances up the feudal hierarchy as self-proclaimed Duke, Prince, King, Emperor, and finally Pope. Like a Frazerian vegetation god he exercises his virility as the seasons revolve. Pretending to be a Duke, he is the lover of the legendary Guinevere, while as Prince he marries the Lady of the Lake, Anaïtis, described as "a vegetation myth of doubtful origin connected with the moon" (137). At the autumn equinox he must leave Anaïtis, compelled by the Master Philologist who considers him a solar hero (a clear satire on Max Müller's school of solar mythology that was in vogue at the beginning of the twentieth century). Jurgen moves to the classically inclined mythical island of Leukê and introducing himself as King of Eubonia he marries a dryad and lives close to Pseudopolis, an ideal city ruled by the masculine and feminine ideals of Achilles and Helen. When Pseudopolis is destroyed by Philistia, a satirical version of conformist and practical American society that turns everything grey and unheroic, Jurgen descends into Hell. Presenting himself as an Emperor he marries a vampire and consorts with Satan's wife. Hell turns out to be a Dantesque construct of his father's mind, while Heaven has been created expressly to uphold Jurgen's grandmother's staunch faith in the Book of Revelation, as Jurgen finds out when he ascends to Heaven as Pope and even for a brief moment sits on God's throne. Jurgen's quest, in other words, takes him through the main cultural influences of Western civilization, from prehistoric fertility rites through the world of classical antiquity and the Age of Chivalry to Christianity, with stress on religious, heroic, and romantic ideals.

It is the jaunty masculine figure of Jurgen that makes this a comic story, for Cabell's cynical and ultimately mediocre middle-aged pawnbroker in the body of a youth subverts the traditional role of the hero and conventional notions of masculinity. Jurgen is an anti-heroic trickster figure, moving onwards and upwards thanks to his cunning and his sexual exploits, rather than heroic deeds of valor. In addition, his masculine powers must be covert, because of the peculiar shadow that follows him and reports back to the Great Goddess, who demands unremitting chastity. Nonetheless, Jurgen manages to trick most of the females he encounters, including manifestations of the Great Goddess, and so thinks of himself as "a monstrous clever fellow" (14), a phrase often repeated throughout the narrative.

Jurgen moves through the world of myth and legend unsurprised by its exotic and metamorphic nature, succeeding on the whole in reducing it to the commonplace and domestic, thus generating bathetic humor. A typical example is Jurgen's complaint to Anaïtis about her "family" of various mythological creatures:

> I cannot congratulate you upon your kindred, for I do not get on at all well with your patchwork combinations, that are one-third man and the other two-thirds a vulgar fraction of bull or hawk or goat or serpent or jackal or what not … Besides, they have no conversation. They merely bellow or low or gibber or purr, according to their respective incarnations—about unspeakable mysteries and monstrous pleasures until I am driven to the verge of virtue by their imbecility. (143, 145)

This is quintessential Cabell—the exotic mating with a goddess figure reduced to cosy domesticity, the literal treatment of mythical figures, and the inversion of accepted social conventions (virtue induced by talk of unspeakable pleasures). When Anaïtis waxes jealous of her guests, Jurgen playfully scolds her:

> My darling, pray consider! Be reasonable about it. Your feminine guests at present are Sekhmet in the form of a lioness, Io incarnated as a cow, Hakt as a frog, Derceto as a sturgeon, and—ah, yes!—Thoueris as a hippopotamus. I leave it to your sense of justice, dear Anaïtis, if of ladies with such tastes in dress a lovely myth like you can be reasonably jealous. (145)

Cabell points out the fictionality of humanity's beliefs and the importance of language in formulating them, hence the stress on wordplay, puns, verbal jokes, elaborate interplay of signifiers. What we are left with as readers is the memory of exuberant verbal constructs of fantastic realms, based on various myths and their prolific commentators, such as Frazer, Müller, and Freud.

Let us concentrate on three such realms, to see how Cabell creates and tranforms them: the garden between dawn and sunrise, Hell and Heaven. We are forewarned that Jurgen is setting out on a satanic or mythical quest when, after following intructions to remove the cross around his neck, he follows an apparition of his wife into an ominous cave on Walpurga's Eve (the eve of May Day), the traditional night of witches' sabbaths, which immediately alerts us to Jurgen as a Faustian figure. There he enters pagan realms and promptly meets a centaur, who takes him to the garden between dawn and sunrise, where dwell the illusions of first love. The centaur explains at the outset that "this garden does not exist, and never did exist, in what men humorously called real life; so that of course only imaginary creatures such as I can enter it" (9). Jurgen reaches the magical garden by riding the centaur as he overtakes the setting sun. There, he meets his first love and relives his illusions about her. First loves live in this mythical realm inhabited by centaurs, Amazons, and other creatures of fantasy, because they are imaginary, the creations of our own minds, ordinary people transformed by love into illusions of perfection.

The centaur mentions that they are going "over the grave of a dream and through the malice of time" (10). And indeed the whole of Jurgen's journey serves to convey to him to what extent his dreams have died and just how maliciously and inevitably time had dealt with him by slowly transforming an eager and idealistic boy in the throes of his first love affair to a tired and cynical petty shopkeeper with barely any pretensions of being a poet left to him. The "malice of time" enacts a deadening metamorphosis on us all. The garden between dawn and sunrise is the place of transitory illusions of first youth and it morphs into a desolate landscape each morning after dawn, paralleling the growing loss of hopes and dreams as we age. It is entirely a place of the mind.

As are all the settings that Jurgen's quest takes him through. The land of Cockaigne, for instance, is a hybrid fantasy place, mixing medieval folktale with Asian fertility rites as expounded in *The Golden Bough* and peopled by the human-animal combination gods of the ancient Mediterranean, particularly Egypt and Asia Minor. Our hero is ousted from it by the power of language, for the Master Philologist has the last word here and can force transmigrations to other mythical places. Max Müller and Sir James Frazer are the patron saints of Jurgen's journey from medieval folklore, through fertility myths to classical Greco-Roman mythology, on to Christian Hell and Heaven. All these imaginary constructs undergo transformations as we read and both fulfill and refute our expectations. Marina Warner, in her fascinating 2001 Clarendon lectures on *Fantastic Metamorphoses, Other Worlds,* focuses on Hieronymus Bosch's triptych, *The Garden of Earthly Delights* (ca.1504) with its post-Ovidian combinations of humans, birds, and vegetation. Warner argues that such exuberant hybridities were stimulated by the mingling of Old World and New in the European discovery of the Americas. One can discern a certain parallel between this painting and Cabell's novel, for the outer wings acting as shutters depict the "Creation of the World," which when opened display "The Garden of Earthly Delights" as the central panel, flanked by "The Garden of Eden" and

"Hell." Cabell's garden between dawn and sunrise represents the Edenic state, with each boy and girl a new Adam and Eve, the land of Cockaigne forms a garden of earthly, particularly sexual, delights, while Hell is the place par excellence of metamorphoses, as Warner points out in her comment: "in the Christian Heaven, nothing is mutable, whereas in hell, everything combines and recombines in terrible amalgams, compounds, breeding hybrids, monsters—and mutants" (36). It is a testimony to the fertility of the human imagination and the mind's attraction to and dread of hybridity.

In Cabell's Hell, Grandfather Satan, a hybrid creature like an oldish man covered in gray fur, with stag's horns and a squirrel's tail, informs Jurgen of the curious origins of Hell: "how he, and all the domain and all the infernal hierarchies he ruled, had been created extempore by Koshchei, to humour the pride of Jurgen's forefathers" (226). Hell is the creation of Jurgen's father, who is so proud and conscious of his sins that he demands extra service from the exhausted devils who cannot keep up with Coth's ideas of torment. Even the lovely vampire Florimel that Jurgen weds may well be an idea that Jurgen provokes his father into conjuring up in his imagination during their acrimonious talk in Hell. The reason Koshchei the Deathless has created Hell and Heaven is to humor and honor the two human traits he cannot possess—pride and love. Heaven is inspired by Steinvor, Jurgen's grandmother, her faith in the Book of Revelation and the power of her love for her family. Thanks to this love, as full of illusions as that of the couples in the garden between dawn and sunrise, Steinvor has installed in Heaven the boy she remembers Jurgen to have been. The youthfully old Jurgen meets this lad as he enters Heaven. "And Jurgen talked with the boy that he had once been, and stood face to face with all that Jurgen had been and was not any longer. And this was the one happening which befell Jurgen that the writer of the tale lacked heart to tell of " (259). In Heaven Jurgen talks to the God of his grandmother and even briefly sits on God's throne, but realizes that he cannot stay in a place in which he does not believe. St. Peter lets him out through a side door. "'For thus,' St. Peter said,' you may return without loss of time to your own illusions'" (271).

In the end Jurgen finds he must return to his old illusions, his old home, and his old life, all of which he finds more comfortable than the mythic and chivalrous quest he has been propelled into. A reluctant Faust who strikes a reverse bargain with the Devil for a return to mediocrity, he renounces his youthfulness and the three beauties Koshchei tempts him with—Guinevere, Anaïtis, and Helen, symbols of faith, desire, and vision—in order to return to a quiet life with his scolding wife, a diminished Odysseus figure who thankfully returns home to his shrewish Penelope, unable to sustain heroic ideals any longer. Offered Queen Helen, Jurgen worships a moment at her feet and then declines, sadly confessing to Koshchei that he no longer desires perfection. "For I am transmuted by time's handling! I shudder at the thought of living day-in and day-out with my vision!" (305). Cabell's narrative subverts the traditional role of the hero, as the rejuvenated Jurgen finds he cannot sustain the demands and possibilities of his youthful body in juxtaposition with his world-weary, aging mind. *Jurgen* is a commentary on art and life from the perspective of middle age, where the poignant and comic coexist, offering ironic distance and surprising insights, particularly about the impossibility of recapturing youthful hopes and ambitions. Made up of a variety of irrealistic materials, mythical and magical, the novel refutes the charge of escapism by painting with great fidelity the movement from youth to middle age and the impossibility of recapturing youthful illusions and hopes. By using such a wide spectrum of traditional materials that transform and morph one into another Cabell is able to create a feeling of history and myth pressing down on his protagonist, a sense of *longue dureé* that weighs heavy on Jurgen and gives him reason to long for home, a comfortable if unexciting place.

Cabell has created here a *Bildungsroman à rebours*, I would argue, an education in how to accept maturity, with its waning of possibilities and loss of illusions. Godshalk has rightly pointed out, in the context of *Life of Manuel*, that Cabell's novels are "cynical works that rely on the tension between the real and the mythic for their vitality. […] The life of their designs relies on the disjunction we all feel between our desires and our actual accomplishments, between our beliefs and our deeds" (121). In *Jurgen* Cabell deploys a remarkable range of metamorphoses of time and place so as to reveal our tendency towards superimposing meaningful patterns on existence; in this way he allows us to both laugh at our capacity for self-delusion and celebrate our creative urges. Metamorphosis here is the primary principle of life—we are all "transmuted by time's handling" and so are the spaces we inhabit, both outside us and, most importantly, within the precincts of our minds.

Works Cited

Cabell, James Branch. *Jurgen: A Comedy of Justice*. New York: Dover, 1977.

Davis, Joe Lee. *James Branch Cabell*. New York: Twayne, 1962.

Godshalk, W.L. "James Branch Cabell: The Life of His Design." *James Branch Cabell: Centennial Essays*. M. Thomas Inge and Egrar E. Macdonald. Baton Rouge: Louisiana State UP, 1983. 108-121.

Patterson, Bill. "The Heir of James Branch Cabell: The Biography of the Life of the Biography of the Life of Manuel (A Comedy of Inheritances)." Winner of the 2000 Cabell Prize. Virginia Commonwealth University Libraries website on James Branch Cabell. 12 December 2007. <http://www.library.vcu.edu/jbc/speccoll/exhibit/cabell/prize3.html>.

Rubin, Louis D., Jr. An Afterword to a New Edition of *Jurgen*. James Branch Cabell, *Jurgen*. New York: Avon, 1964.

Tarrant, Desmond. *James Branch Cabell: The Dream and the Reality*. Norman: University of Oklahoma Press, 1967.

Warner, Marina. *Fantastic Metamorphoses, Other Worlds: Ways of Telling the Self*. The Clarendon Lectures in English 2001. Oxford: Oxford UP, 2002.

Works Consulted

Massey, Irving. *The Gaping Pig: Literature and Metamorphosis*. Berkeley: University of California Press, 1976.

Skulsky, Harold. *Metamorphosis: The Mind in Exile*. Cambridge, MA: Harvard UP, 1981.

PLACES IN METAMORPHOSIS AND REPRESENTATIONS OF AGING IN BRAZIL

GUITA GRIN DERBERT

Abstract

This paper discusses the metamorphosis of identities created by new images of aging and gerontological discourse in a Brazilian context. The central argument is that these images and discourses transform old age into a problem of individual neglect, a problem of failed consumers who were unable to adopt lifestyles and forms of consumption capable of assuring their well-being. Three intertwined social processes give a special configuration to this sense of the place of the aging other in Brazil: 1) new patterns of retirement that include an ever younger segment of the population among the retired people, 2) the creation of a series of new stages that separate adult life from old age such as "middle age," "third age," and "active retirement," 3) new body technologies that transform Brazil into one of the leading countries as far as esthetic plastic surgery is concerned. A variety of Brazilian textual and ethnographic materials are examined to explore local disjunction of class and gender related to these social processes: gerontological literature produced by the Brazilian Gerontological Society, representations of aging and practices carried out by professionals and participants in recreational and educational programs for the third age, and images of aging in the media.

> "Bad moods and depression, common among the elderly, are the consequences, not the causes of a badly-run life."

> "The important thing is to prevent. The geriatrician is not a doctor who specializes in treating worn-out old people. [...] Actually, the leading edge in geriatrics is prevention, which involves some basic recommendations. A positive attitude towards life, for example, is essential."

> "Another important tip is doing exercise. [...] Keeping fit reduces the degeneration of the neurotransmitters, responsible for directing information through the nervous system, and reduces the risk of mental disorders such as schizophrenia."

> "Miracles, however, do not exist. [...] a good doctor is going to recommend exercise and regular cholesterol exams, not vitamins [...] since youth in capsules has not been invented yet, and probably never will be. This means that whoever wants to enjoy old age is going to have to work at it."

These are excerpts from interviews with Brazil's best-known geriatricians for a report published in a popular Brazilian magazine (*Globo Ciência*, no. 30, 2002). They illustrate the central argument of this essay on the metamorphosis time operates on the aging body: old age is transformed into a problem of failed consumers that were unable to adopt lifestyles and forms of consumption capable of assuring their youth.

Brazil is a country in which the cult of beauty, youth, and sensuality has historically been considered the most prominent cultural characteristic along the axes of race, class, and ethnicity. This extreme emphasis on youth has as its corollary a premature aging of the population. Since the eighteenth century, travelers and historians have contrasted the beauty and sensuality of the young women with what Gilberto Freyre, one of the first Brazilian anthropologists, called the "negligence of body of our matrons above the age of eighteen" (Cited in Freyre 363-64).

In 1777, in her *Letters from the Islands of Teneriff, Brazil, the Cape of Good Hope and the East Indies,* Mrs. Kindersley commented that the elite women in Brazil "look[ed] old very early in life," and quickly lost the delicacy and charm that had characterized them. Mrs. Maria Graham, in her *Journal of a Voyage to Brazil and Residence there during the years 1821, 1822, 1823,* was more vehement when she observed that elite women in the northeastern state of Bahia became "almost indecently slovenly after very early youth." Similarly, in discussing Rio de Janeiro in the first decades of the nineteenth century, John Luccock said that among elite women, at eighteen they are already matrons, having attained complete maturity: "After twenty a decline. They become fat and flabby, develop a double chin, turn pale, or else they dry and wither [...] they grow more ugly, with down on their faces and with the air of a man or of a virago" (Cited in Freyre 363-364).

Premature aging was also a characteristic of Brazilian men. Kidder and Fletcher described elite young men, in the second half of the nineteenth century, in the following terms:

> He is made a little old man before he is twelve years of age, - having his stiff black silk-hat, standing collar, and cane; and in the city he walks along as if everybody were looking at him, and as if he were encased in corsets. He does not run, or jump, or trundle hoop, or throw stones as boys in Europe and North America (Cited in Freyre 404).

Until the middle of the twentieth century, aging was an essentially private question, involving the family, individual planning, or charitable organizations. Other than the military, very few institutions could guarantee a retirement pension for their employees.

As of the 1980s, international organizations like the United Nations played an active role in developing new approaches to aging in different countries. These organizations pressured national governments to promote specific public policies about their older populations and created criteria for evaluating the results.[1] This global orientation rearticulated the meanings given to the most advanced stage of life in different cultural contexts.

In Brazil, the metamorphosis of old age from a private concern into a field of political practices preceded these initiatives on a global scale and became most evident in the debates on retirement pension levels and the policies for financing them. In the 1930s retirement was transformed into a right of the urban worker and was granted in relation to the years worked (25 years for women, and 30 years for men) and not by chronological age.[2] For this reason, the age of retirees today is relatively premature; in 2001, more than half of the 23 million Brazilian retirees were under 50 years of age.

In the 1980s, the right to retirement was extended to rural workers and it was only in the 1990s that the right to retirement was made universal in the sense that the equivalent of one minimum monthly salary was granted to all Brazilians 70 years of age and older without other resources for survival.[3] The disparity in retirement wages and pensions of employees from different occupational sectors – such as judges or members of congress, who may receive benefits exceeding one hundred minimum

---

[1] For the United Nation Conference on Aging see Cohen.

[2] Retirement pensions are paid by the state in exchange for a monthly contribution made by the worker and his or her company. The predominant model in Brazil is what is conventionally called the German Bismark model in which the retirement pension system is organized on an occupational basis.

[3] Wages are set according to multiples of the federally mandated monthly minimum salary, such that someone might express their earnings in terms of making three minimum salaries per month, or ten, or forty. This measure is also the basis for retirement benefits.

salaries from the state – is a pressing issue in current political debates about social and economic injustices in Brazilian society. In spite of being a country with large expanses of land, 180 million inhabitants and the eighth largest economy in the world, Brazil has one of the highest indices of social inequality, where about 29% of the population lives on less than US$1.00 per day.

It is in this context of premature aging and retirement that another type of metamorphosis of old age takes place. This essay discusses the way in which the "third age" operates as a way of resurrection and celebration of life and as such erases old age from Brazilian social concerns. To explore the set of signifying practices this category articulates, I describe the arenas of ethical dilemmas shaped by different actors interested in designing a positive experience of aging: gerontologists, organizations for the third age, and the media.

The new generation of Brazilian gerontologists enthusiastically combats images of old age as a period of physical decline and absence of social roles. Inspired by international gerontology, they emphasize the advantages brought by aging and were active in promoting organizations where a successful aging experience can be shared. These recreational spaces linked to both governmental and non-governmental organizations include different types of programmed activities such as senior citizen centers, open schools for the "best age," and universities for the third age. The initial goal of these initiatives was to bring together people over 70 years of age; instead, they mobilize mostly women in their 50s or 60s who are interested in making aging a radically different experience from that of women in previous generations. These recreational groups are spaces for enthusiastically refuting old age by transforming the third age into an opportunity to discover new identities, carry out projects abandoned in earlier stages of life, and establish more rewarding relationships with people in their or in other age groups.

The interaction between experts, organizations for the third age, and the media allows healthy people with autonomy and sufficient income to experience a more gratifying aging process, which offers these young old people opportunities to renew their bodies, identities, and self-image. In this sense, contemporary Brazil appears to be quite distant from that place in which matrons become "almost indecently slovenly after very early youth" (Graham).

Nevertheless, this commitment to positive aging conceals the specific problems of old age. The loss of cognitive abilities and physical and emotional control – abilities that in democratic societies are fundamental for the recognition of the individual as an autonomous person capable of fully exercising the rights of citizenship – are seen as the result of transgressions committed by individuals against their own bodies and health. In a country that combines blatant social hierarchies with a consumer culture, old age is transformed into a problem of failed consumers who have been unable to choose goods and services that may forestall aging. Old age may thus vanish once again from the range of wider Brazilian social concerns.

The Uses of the Third Age

The third age is an expression which was recently introduced and quickly absorbed into the Brazilian vocabulary.[4] Its current use among researchers interested in the study of aging does not make reference to a precise chronological age, but instead is a term for referring to older people that has not yet acquired a disparaging connotation. The invention of the third age is understood by authors such as Guillemard (1986) and Lenoir (1979) as the result of the universalization of the right to retirement pension guaranteeing that the last stage of life would correspond to remunerated inactivity. Since the 1970s, the elderly – seen as the most underprivileged sector of European society in the 1940s and 1950s – could no longer be considered a segment of the population that is deprived of economic resources. The third age is not synonymous with decay, poverty, and illness, but is instead seen as a time to enjoy activities while free from the constraints of professional life.

In Brazil, the universalization of retirement and old age pensions guarantees the elderly economic conditions not afforded to the population in other age groups, especially among youths. Studies have shown that the average income of those over 60 years of age is greater than that of those under 30 years of age, and that in families which include elderly people, 52% of the family income comes from their income. In these families, 49% were headed by an elderly person and had children living in the same household (Camarano 1999). Thus research on dependence between generations challenges the idea that the elderly are a burden for the family. On the contrary, the presence of a retiree in a home is one of the reasons for a lower rate of poverty; in other words, poverty among the younger

---

[4] The expression originated in France in the 1970s with the creation of the *Universités du Troisème Age*, and was then incorporated into British vocabulary with the creation of the Universities of the Third Age in Cambridge, England, in the summer of 1981 (Laslett).

generations would be two percentage points greater if the elderly person lived in a separate home (Barros et al. 1999).

The privileged position of the elderly in relation to the other segments of the population revealed by the quantitative data should not mislead us with regard to the gross differences in income within this segment of the population, but these data show that in all classes the elderly enjoy greater consumer power than youths.

Premature retirement and the fact that retirement does not mark the transition of individuals into old age establish a peculiar relation with the way in which a series of intermediate stages of adult life are created, such as "middle age," the "third age," and "active retirement." These new stages of adult life are not perceived as mature interludes that precede old age, but are propitious moments to realize dreams deferred in other stages of life and to enthusiastically combat aging symptoms.

Dona Lázara, a 62-year-old seamstress who participated in a recreational group for the third age doing dance, singing, and yoga can be considered an icon of the way these new images of aging are promoted when she made the following comment:

> My big transformation began when my daughter gave me an unexpected gift: she signed me up for a course in body expression. I was 49 years old and had never worn shorts. In the first classes I felt ridiculous, old, awkward. I continued and my mind opened – it opened to the world. I lost weight and began to feel alive (*Claudia*).

The third age does not have as a reference a specific socio-economic stratum, it is, primarily, a call to action for those who feel prematurely aged to engage in a struggle against attitudes and behaviors stereotypically related to old age.

In Brazil, a new language actively deconstructs age as a relevant marker for behavior and lifestyle. This deconstruction is certainly fed by consumer society's conceptions of the body and health. According to Featherstone,

> Consumer culture latches onto the prevalent self-preservationist conception of the body, which encourages the individual to adopt instrumental strategies to combat the deterioration and decay, (applauded too by the state bureaucracy who seek to reduce health costs by educating the public against body neglect) and combines it with the notion that the body is a vehicle of pleasure and self-expression (170).

The view that the body is entirely moldable and that its imperfections are neither natural nor immutable is radicalized in a context in which the cults of beauty and sensuality are set against the premature aging of the population.

A new market of goods and services emerges in which flaccidity, wrinkles, and even dissatisfaction and sadness are transformed into indicators of some sort of moral lassitude that should be treated with the help of the cosmetics industry, physical exercise, and the motivational activities offered through the recreational centers for the third age.

The growing interest in Brazilian society for rejuvenation technologies casts these new images of aging into sharp relief. According to the Brazilian Society of Plastic Surgery, in the year 2000 Brazil was the country in which the greatest number of plastic surgeries for aesthetic purposes was performed. Nearly 350,000 people underwent at least one aesthetic surgery, a number equivalent to 207 out of every 100,000 inhabitants of the country. For the same year, the United States, the historical leader in this ranking, registered a rate of 185 individuals who underwent aesthetic surgery for every 100,000 inhabitants.

The struggle against aging is the main reason given for aesthetic plastic surgery, followed by the desire to correct physical imperfections because of their psychological burden, and finally an interest in the sculpting of bodies. These data show that this kind of surgical intervention is not the exclusive privilege of small economic elites, but is also sought by members of the middle classes. It is mostly women who invest in this type of surgery, but men underwent 30% of the total number of aesthetic plastic surgeries carried out in 2000, while 13% of the total number was conducted on minors under the age of 18 ("*Brasil, império do bisturi*").

The plastic surgery market reaches ever lower income levels in the population. By parceling the payments for surgery, which may be extended over as much as a 24-month period, these interventions are no longer the privilege of the most wealthy. The plastic surgeon Pitanguy, for example, is so popular in the country that he was the theme of a samba song during *Carnaval* of 1999.[5] Accordingly, exercise gym centers are part of the landscape of the poorest neighborhoods in large Brazilian cities.[6] In an analysis of body fitness among the inhabitants of Rio de Janeiro, anthropologist Goldenberg asserted that "there are no fat or ugly people, only lazy people."

The transformation of the most advanced life stages into privileged moments for self-realization, the fact that most elderly in all social classes tend to have greater resources for consumption, and the self-preserving

[5] "In the universe of beauty," Master Pitanguy – Carnival song, by the Rio de Janeiro samba school Caprichosos de Pilares.

[6] On gyms in Rio de Janeiro, see Malysse.

conceptions of the body all give new meaning to the experience of aging in Brazil, transforming the third age into a form of negation of old age and the stereotypes associated with it.

The visibility that the third age has attained in Brazil led to reformulations of the gerontological discourse whereby the perspective of misery in the treatment of old age was replaced with a view of the elderly as a source of resources.

Gerontological Discourse and New Images of Aging

The new representations of aging contrast with the images of aging produced by the Brazilian gerontological discourse. Since the 1960s, the gerontologists who founded the first Brazilian geriatrics and gerontological association have been active in transforming old age into a social concern. In addition to the concern of combating charlatanism and offering standardized scientific training for the treatment of old age, a basic role of the association is the attempt to convince the state that old age deserves specific public policies.

The main strategy use to this end was to construct an image of older Brazilians as a miserable and suffering population. A 1991 manual of social gerontology illustrates the type of image produced:

> [...] the older person is someone who is discriminated against, inactive, lives in precarious conditions, who has lost status and prestige. [...] Elderly people live in crisis, an identity crisis [...] characterized by isolation, loneliness, disinterest in life, alcoholism, decrepitude, senility, and social and physical death (Sá, 19).

However, since the 1980s the image of the older person as a living in misery has come to coexist with the perspective of the elderly as a resource, which emphasizes the advantages brought by aging. This new perspective was stimulated by the World Assembly on Aging, organized by the United Nations, that declared 1982, the "International Year of the Aged."[7] An International Plan was then proposed that defined the old person as an integrated whole, in need of special attention, and that at the same time seeks to rediscover his or her place in society and recover his or her self-esteem.

The representation of the older person as a potential resource was also a response to research results and ethnographic accounts that indicates that

[7] On the effects of this World Assembly in India, see L. Cohen (1998), and in Brazil, see L. Prata.

older people project a much more positive image of their situation than that exposed by gerontological theory.

It became evident that what was known about the elderly in Brazilian gerontology was quite vague, laden with pessimism, and that gerontology fed myths and negative stereotypes of old age. As such, stereotypes of abandonment and solitude were replaced by the image of the elderly as active people, capable of creatively responding to the changes that characterize old age.

This new perspective constitutes an important element to the conversion of programs for the third age into the main public policy on old age. In other words, prior to the 1990s the Brazilian government's policies for the older segment of the population were largely restricted to homes for the elderly and the payment of retirement and pensions. From this point on, public investments became mostly concentrated on programs for the third age. To implement or to expand existing programs for the third age is part of the plan undertaken by most of the public administrations of Brazilian cities, including very poor municipalities in which the population over 60 is relatively small.

With different denomination – as for example conviviality groups, open schools, senior centers, and universities for the third age – these programs also carried out by non governmental organizations are the most frequent illustration employed by gerontologists to demonstrate that the experience of aging can be an innovative and gratifying experience.

The Aging Experience in Programs for the Third Age

> "I think the third age is an innovation, the best possible one about the elderly. It is the best because is gives us total freedom, you know. It's delightful."

> "I am here of my own free will. I always liked studying and I am taking advantage of an opportunity. I am happy, I love coming here, the atmosphere is so good, and I like sharing experience with my peers. All of this is a pleasure for me."

> "I felt, even at my age, that I could still be somebody, always fighting the fight."

These statements by women in their 60s in a university for the third age in São Paulo show the enthusiasm with which participants describe their experiences in the different programs for the third age. Open to all older people, each organization mobilizes people with different socioeconomic profiles. For example, the universities for the third age draw a segment of the population with greater consumer power and the conviviality groups organized by municipal governments or churches are directed towards the very poor. In spite of the great diversity of human resources and equipment available in each setting, the proposed activities are similar. In every program there are multiple activities, including dance, chorus, tourism, and projects involving memory, autobiography, and reminiscences. Crafts are more common in the poorer programs, while courses and lectures are offered more frequently at the universities. However, all of the programs are highly interested in promoting the self-esteem of the participants and challenging the stereotypes and prejudices about the elderly.[8]

The construction of a positive image of aging among the students does not refer to the elderly as repositories of wisdom or experience. It is primarily the readiness to learn and gain new experiences that gives this group its identity and uniqueness making the participants feel radically different from older people.

The programs offer a space in which reshaping traditional patterns of aging can be a collective experience, and active participation in these programs is a way to intensely experience a new stage in life, a moment for the exploration of new identities and new forms of self-expression.

The Media and the Third Age as a New Consumer Market

The celebration of the third age is also evident in other manifestations such as the women's magazines for the middle and upper classes. In an analysis of these periodicals, Pires shows that distinct images of aging are being produced. In the contexts dealing with beauty and fitness, the magazines tend to call on all women, even those who have not yet reached adulthood, to engage in a battle against aging, offering a range of goods and services capable of preventing or retarding aging. However, in the contexts in which the elderly or old age is addressed as a real situation, it is shown as a rewarding period for individual self-realization. Readers are shown older men and women who have found a new professional career later in life or engaged in new activities, realizing their youthful dreams that were deferred because of the obligations of adult life. Women, in particular, through physical fitness activities, like dance and exercise, rediscover their own bodies and feel rejuvenated.

[8] On the programs for the third age see Barros (1998), Cabral (1997), Debert (1999), Eckert, Guerrero, Mattos, Motta, Peixoto and Prata.

The representations of the elderly in these magazines are of active people who were capable of finding new and interesting activities at this stage of life. Old age is the moment in which a woman, freed from the social roles of earlier stages of life, can finally dedicate herself to personal fulfillment. The magazines create a new actor, defining a new consumer market in which the promise of eternal youth is the subtext through which a new wardrobe, new forms of leisure, a new relationship with the body, and with family and friends are offered. There is no room for images of illness, physical decay or dependence as the destiny of those who grow older. More than presenting the last stage of life, they show to those who do not feel old how they should perform. It is also part of media coverage to invite gerontologists and especially geriatric doctors to discuss the scientific advances in the treatment of aging or to suggest simple, inexpensive measures for combating the aging process as shown in the excerpts that open this article.

New Hierarchies and the Reprivatization of Aging

Gerontologists portrayed the elderly as the special victims of poverty. However, the elderly who are researched and depicted in the media are active, lucid, and ready for living one of the happiest periods in life in which the only obligation is personal fulfillment. The view of the misery of old age was certainly fundamental in transforming the elderly into political actors and making Brazilian society more sensitive to problems related to aging and retirement. The fact that the publicly visible elderly are not the neediest ones creates an impasse for gerontologists. While they have intended to serve the needs of one of most fragile sectors of a society, this is not the profile of the clientele that they have mobilized or that which has gained media attention.

In a country in which the population ages prematurely, these young old people seek to revert to old age stereotypes and to adopt consumption patterns and life styles that make aging a gratifying experience.

The success of programs for the third age and the interest in rejuvenation technologies have led gerontologists to deny their own object of research and professional exercise, as can be seen in the quotes at the beginning of this chapter. From being active participants in the constitution of the elderly into political actors they are transformed - especially by the media - into promoters of a panoply of prescriptions for showing those who do not wish to be old how they should act.

In Brazil, the enthusiasm for the programs for the third age overshadows the precariousness of public policies for advanced old age.

The focus on the "young" old and in the programs for the third age occurred at the expense of the very old and those who have lost the functional autonomy necessary to participate in these programs.

It is very difficult to know whether old age will remain an unpleasant secret, which we find increasingly sophisticated ways of denying, as Elias has suggested. It is also possible, as Featherstone has suggested that alternative routes for addressing it may emerge. Hopes that scientific advances may offer solutions to the aging of human cells or that technology will find ways of minimizing the problems of dependency in old age have grown more concrete.

The growth in the number of retirees in Brazil was also accompanied by the increase in their political power and their capacity to demand more and to implement political demands.

Will the elderly be seen as a monstrous verminus bug that greedily devours both the inheritances of younger family members and public resources that should be distributed to other sectors of society? What we do know is that the intergenerational solidarity was a fundamental key in the constitution of modern states.

Works Cited

Barros, R. P. et al. "Incidência e Natureza da pobreza no Brasil." Camarano, A. A., ed. *Muito Além dos 60 – Os Novos Idosos Brasileiros*. Rio de Janeiro: IPEA, 1999.

Barros, M.M.L. *Velhice ou Terceira Idade?* Rio de Janeiro: Editora Fundação Getúlio Vargas, 1998.

"Brasil, império do bisturi," *Veja* 1683, 84-89, Jan. 17, 2001.

Cabral, B. E. S. L. *De Trabalhadores a Aposentados do PRORURAL. As Contradições da Política Social e a Concessão Tardia de Direitos*. Campina Grande: UFPb, 1986.

—. "A Vida Começa Todo Dia." *Revista Estudos Feministas* 5 (1997): 159-168.

Camarano, A. A. *Muito Além dos 60 – Os Novos Idosos Brasileiros*. Rio de Janeiro: IPEA, 1999.

Claudia, No.114, Dec. 1991:110.

Cohen, L. "Não há velhice na Índia: os usos da gerontologia." Debert, G. G., ed. *Antropologia e Velhice*. Campinas: IFCH, UNICAMP, 1994.

—. *No Aging in India: Alzheimer's, the Bad Family, and Other Modern Thing*. Berkeley: University of California Press, 1998.

Debert, G. G. *A Reinvenção da Velhice*. São Paulo: EDUSP, 1999.

—. "La viellesse dans la publicité brésilienne. Une étude des représentations." *Cahiers du Brésil Contemporain* 55/56 (2004): 141-156.

Eckert, C. "A vida em outro ritmo" Barros, M.M.L., ed. *Velhice ou Terceira Idade?* Rio de Janeiro: Editora Fundação Getúlio Vargas, 1998.

Edmonds, A. "No universo da beleza: notas de campo sobre a cirurgia plástica no Rio de Janeiro" Goldenberg, M., ed. *Nu e vestido: dez antropólogos revelam a cultura do corpo carioca.* Rio de Janeiro: Record, 2002.

Elias, N. *Loneliness of the Dying.* Oxford: Blackwell, 1985.

Featherstone, M. "The Body in Consumer Culture" Featherstone, M., ed. *The Body - Social Process and Cultural Theory.* London: Sage, 1992.

Freyre, G. *The Masters and the Slaves.* Berkeley: University of California Press, 1986.

Goldenberg, M. and Ramos, M. S. "A Civilização das formas: o corpo como valor" Goldenberg, M., ed. *Nu e vestido: dez antropólogos revelam a cultura do corpo carioca.* Rio de Janeiro: Record, 2002.

Guerrero, P. *Experiência de Envelhecimento e a Universidade para a Terceira Idade da PUCCAMP.* Campinas: IFCH-UNICAMP, 1993.

Guillemard, A. M. *La Vieillese et L'État.* Paris: PUF, 1986.

Kindersley. *Letters from the Island of Teneriff, Brazil, the Cape of Good Hope and the East Indies.* London: J. Nourse, 1777.

Laslett, P. "The Emergence of the Third Age." *Ageing and Society* 7 (1987): 133-160.

Le Lenoir, R. "L'Invention du "Troisième Age": Constitution du Champ des Agents de Gestion de la Vieillesse." *Actes de la Recherche en Science Sociales* 26 (1979): 57-82.

Malysse, S. "Em busca dos (H)alteres-ego: Olhares franceses nos bastidores da corpolatria carioca" Goldenberg, M. *Nu e vestido: dez antropólogos revelam a cultura do corpo carioca.* Rio de Janeiro: Record, 2002.

Mattos, F. M. *Velha é a Vovozinha - Identidade Feminina na Velhice.* Santa Cruz do Sul: EDUNISC, 1990.

Motta, A. B. "Chegando para a Idade" Barros, M.M.L. *Velhice ou Terceira Idade?* Rio de Janeiro: Editora Fundação Getúlio Vargas, 1998.

Prata, L. E. "Os Programas Especificamente Destinados à População Idosa" SEADE, Fundação. *O idoso na grande São Paulo.* São Paulo: Fundação Sistema Estadual de Análise de Dados (SEADE), 1990.

Pires, A. *Velhos em Revista - envelhecimento e velhice nas páginas de* Cláudia *e* Playboy *(anos 80 e 90)*. Campinas: IFCH-UNICAMP, 1998

Peixoto, C. "Entre o estigma e a compaixão e os termos classificatórios: velho, velhote, idoso e terceira idade" Barros, M.M.L., ed. *Velhice ou Terceira Idade?* Rio de Janeiro: Editora Fundação Getúlio Vargas, 1998.

Sá, J. L. M. *A Universidade da Terceira Idade na PUCCAMP - Proposta e Ação Inicial. Campinas*. Campinas: PUC-Campinas, 1991.

Simões, J. A. "A maior categoria do país. O Aposentado como Ator Político." Barros, M.M.L., ed. *Velhice ou Terceira Idade?* Rio de Janeiro: Editora Fundação Getúlio Vargas, 1998.

METAMORPHOSIS OF SACRAL PLACES IN SOFIA

EKATERINA PANTCHEVA

Abstract

What mechanisms allow churches to later function as mosques, and mosques likewise as churches? What factors lead to reuse (instead of destruction) of the same temple? And what precisely is the metamorphosis they undergo? How can its signs be defined? Can a sacred place be "inherited?" To what extent is its metamorphosis accepted by the different social strata and religious/ethnic groups who make use of that place (not only worshippers, but also academics, visitors or locals)? Non-functioning churches are still accepted as churches by Christians, just as former mosques are perceived as mosques by Muslims. Despite all the transformations, a given locus remains unchanged in its consideration as numinous in different people's perception. Sofia's history, geographico-cultural position and ethno-religious mix make it an intriguing case study.

When one culture comes to dominate another, it adopts or adapts its predecessor's spiritual sites. This process of making the "foreign" into one's "own" is manifested in various ways: destruction, prohibition, desacralisation or even deliberately inappropriate use. Sometimes, however, some sacred buildings survive as sacred even after transformation. Such places are what interests me most in this paper.

The examples I shall focus on are from my hometown, Sofia. In the sixteenth century, after the Ottoman conquest, some originally Orthodox churches were converted into mosques – the basilica of St. Sofia and the rotunda of St. George – and back into churches after the Liberation. Conversely, the church of the Seven Saints there was originally founded as a mosque in the same century.

All three sites are crucial for the town's history and its self-image. The rotunda is the oldest temple in the city, the basilica gave it its present name – Sofia – and the church of the Seven Saints is on one of the two spots where scholars have hypothesised that the relics of our national saint Ivan Rilski rested after being brought from Turnovo in the eleventh century.

The first question I would like to raise is: What mechanisms allow churches to later function as mosques, and mosques likewise as churches? Here I am not speaking of the purely theological aspect. Rather, about the way in which a people or a community acknowledges that a certain building is indeed what its title suggests that it is – in fact as well as in name. And I am specifically talking about how that assurance will not be shaken by later metamorphoses.

The metamorphoses of these sacral buildings should give us some evidence about the patterns of belief and the attitudes shown towards such sites. The Koca Mehmed Pasha Mosque, popularly known as the Black Mosque, was erected in 1528 on the site of a former nunnery of the Rila Monastery. In 1901, when the transformation of the mosque into the church of the Seven Saints was undertaken, the ruins of an early Christian church were excavated, together with an ancient pagan temple dedicated to Asclepius. The other two churches – of St. George and of St. Sofia – were also founded on places perceived as sacral from Antiquity onwards. Under the present basilica of St. Sofia there are the remains of four earlier Christian temples, the earliest of which dates back to the fourth century (and in it the Council of 343 took place).

Of course, it is a matter of common knowledge that pagan places of worship were often transformed into temples for adherents of the three Abrahamic faiths, the "peoples of the Book," as they are known. In such cases the site had to be cleansed, purged, exorcised of association with the pagan deities. Frequently nothing less than destruction would do. But there is a principal difference in the metamorphosis of a pagan sacred place into a Christian one and of a Christian temple into a mosque (or indeed vice versa). The first case involves demolition, which is not compulsory in the second (or the third). However, transforming a church into a mosque does seem to involve a tremendous humiliation for the locals, especially when a central church is concerned. Sometimes there is no worse humiliation than forced metamorphosis – here, conquering the soul of a city by transforming it into a foreign sacred place.

Concerning the factors to reuse (instead of destruction) of the same temple, four types could be distinguished, namely practical, strategic, political and artistic. The practical factors most often mean that for reasons of time and finance, any conqueror cannot afford to avoid economy of effort. When incorporating a city or state into what for the conquered people is an alien commonwealth (*oikoumene*), whether that incorporation is pursued through force or persuasion - why should the conqueror cut off his nose to spite his face? It is always easier to make use of objects close at hand. In the case of the rotunda in particular, the very shape of the building seems to be quite proper for a mosque, as it was a round dome construction built on a square base. So aesthetics could be maintained with the minimum of effort.

As far as strategic factors go, it is well-known that St. Sofia was initially the biggest church in the town; besides, it was located on the highest part of it and was also a metropolitan church. Its fame was so great that the whole town was associated with it and as early as the fourteenth century we have evidence that Sredets was by then already called Sofia (from the charter of 1382 given by Tsar Ivan Shishman to the Monastery of Dragalevtsi). So its prestige must have brought it to the Ottomans' attention. In the case of the Black Mosque we know that it had one of the biggest domes in the Balkans – so is this too then a strategic reason, in the psychological sense – a matter of prestige, just as with St. Sofia centuries earlier?

An example of political or ethical factors is the opposite process of metamorphosis, Muslim to Christian. There is written evidence, even though a dubious one, that at the time after the Liberation, the Turks used to leave their mosques with all the estates attached under the condition that they were to be turned into churches so that their sacrality is preserved. Well, why should they not do so? A member of one confession may indeed appreciate the ethics of another confession – and transforming a temple of one stripe into a temple of another may be seen as less blasphemous than transforming it into a place of oppression or even commerce. Some Muslims may have been sad to see the Black Mosque taken from them – but transformation into a church may have been the lesser of innumerable evils.

Yet were there artistic factors as well? If it was a significant and beautiful building – as was probably the case with the Black Mosque – why not preserve it? Again, even in the case of some pagan temples, aesthetics triumph: better to have a Parthenon or a Stonehenge than a supermarket, and if the preferred choice is a religious building of another faith, then there is no reason to destroy the original form along with the function. However, certain aspects may trouble its new possessors – what today we do through censorship, the Ottomans did with whitewash and plaster, covering up – not always destroying – those images that contravened the rules of Islam.

Now, what factors influence the secularisation of such buildings? After the Liberation, the building of the Black Mosque was firstly used as a storehouse (just like the majority of the surviving Turkish buildings) – as a storage for weapons. Later a leading political figure, Stefan Stambolov, ordered that the buildings related to the mosque were to be turned into a prison. Here we might posit purely pragmatic reasons. In a time of struggle almost anything may be placed at the service of the people who are struggling. A defining landmark may be used in this way. So indeed might a people's most sacred places. Why then, should not the outbuildings of a temple of another faith be used in this way? This is not blasphemy or mockery but *Realpolitik*.

After the earthquake that destroyed St. Sofia's minaret in 1858, the mosque was abandoned and used as a warehouse until the Liberation in 1878. Then this secular trend was taken by the Bulgarians and the building was used to store the gas for the town's street lamps. Besides, what was built on its dome was a tower from which the fire brigade could observe the whole area for fires.

Let me move to the question of what precisely the signs of the metamorphosis are. I will start with the signs pretending to signify the conversion of churches into mosques. To begin with, some of the changes aim solely at reconstruction. In the case of St. Sofia, for instance, very tiny bricks were attached to the old dilapidated walls (like some sort of tiling). Besides this, during the Ottoman times there were other construction changes, for example the blocking-up of windows and entrances both in the rotunda and the basilica. The blocking of some windows might have been for religious – which is to say iconographical – reasons, as often the combination of windows was in the form of a cross. The most serious change in St. Sofia involved the removal of the first eastern column of the southern arcade and the making of a connection between the neighbouring columns, in the form of a peaked stone arch, with the view to adapting this sector for the vizier. The inscribed cross of St. Sofia – being a three-nave basilica with one transept – was changed into another form by adding a minaret to it. Among the decorative signs there are also not purely decorative ones, such as the covering of the murals with painted floral motifs. It should be mentioned, however, that it was done in a very careful way without damaging the murals below it. The most crucial change signifying the turning of a church into a mosque was the minaret and – in the case of St. Sofia – a niche facing Mecca. Last but not least, it was the change of the name.

In the case of transforming the mosque into a church, the precise signs of the metamorphosis are: the addition of an altar, of two entrance side niches and of a narthex, so that the cross-like form is achieved and the main construction of the former square mosque and its dome – sanctified. Furthermore, a bell tower was erected. So the construction changes here were much more serious than in the case of the churches turned into mosques. After being depicted from the inside, the church was inaugurated on July 27 1903 – on the patron saints' day and named after the Seven Saints. All the buildings attached to the mosque were removed.

We need to look at conscious acclimatisation. To what extent is a site's metamorphosis accepted by the different social strata and religious/ethnic groups who make use of that place (not only worshippers, but also academics, visitors or locals)? Non-functioning churches are still accepted as churches by Christians, just as former mosques are perceived as mosques by Muslims. Despite all the transformations, a given locus remains unchanged in its consideration as numinous in different people's perception. So, are reuse on the one hand, and destruction (with "replacement" by a functionally different religious building) on the same site, on the other, kinds of metamorphosis? If so, are they fundamentally different processes? How does metamorphosis affect the perception of holiness? And how is this reflected in secondary sources? For the purposes of this article, the secondary sources are legends – tales of a building's history. They are not apocryphal but anecdotal.

Let us look at these churches in turn. Sinan, the architect who built the Black Mosque (later to be transformed into the church of the Seven Saints), was born into a Christian family. There are legends according to which he was a Bulgarian (sometimes associated with the mythological figure of Master Manol) and a scholarly study also claims it was the case (Shishmanov). According to other sources, he was a Serbian, or a Greek, or an Armenian janissary.

In the case of St. Sofia, the second big earthquake in 1858 destroyed its narthex and, again, the minaret (even though not completely as it was by the earthquake in 1818). According to a Bulgarian legend this happened on a Friday, just before the hour of the main prayer. Inside were only the two sons of the imam, and they died. This, so we are told, was an omen for the Muslims, and they abandoned the sacred building, using just the southern part as a warehouse.

Herein lies the crux of my paper. Perhaps, in those turbulent times, a process of anthropomorphism and "transference of epithet" took place in the folk memory. The very real phenomenon of resistance on the part of a people may have been attributed also to inanimate objects – buildings, or the very land. The site itself was seen to resist metamorphosis, assimilation, or co-opting.

So was it hypocritical – or at the very least ingenuous – of the Bulgarians to turn a mosque into a church, if (as we assume) they were unaware that it had indeed once been the site of a Christian temple? Perhaps not. Coincidence alone will not acquit them of such a charge – but there may be a factor that will. Certainly all the practical factors outlined here may have played a part.

Works Cited

Bojadzhiev, Stefan. *Sofijskata tsărkvata "Sv. Sofija"* [The Church of St Sofia in Sofia]. Sofia: Bălgarski Khudozhnik, 1967.

Dobrev, Ivan. *Sv Ivan Rilski.* [St Ivan Rilski] Altbulgarische Studien: Beiträge zur Kultur-, Literatur-, Sprach und Kunstgeschichte Bulgariens. Herausgegeben von Dimitar Dunkov. Band 6. Linz: Slavia Verlag, 2007.

Filov, Bogdan. *Sofijskata tsărkva "Sv. Sofija". Sofijskata tsărkva "Sv. Georgi".* [The Church of St Sofia in Sofia. The Church of St George in Sofia.] Sofia: Iztok – Zapad, 2004.

Ilieva, Sonya. "Mythology as an Alternative for Motion of Created World." [*http://thracology.dir.bg/english/myth_en.html*]. Accessed 2007-07-12.

Karamihaleva, Aleksandra. 100-godishninata na hrama "Sveti Sedmochislenitsi" [100th Anniversary of the Temple of Seven Saints]. *Tsarkoven vestnik* 2003-06-15.

Miteva, Zhanet. Hramăt Sveti Sedmochislenici [The Temple of the Seven Saints]. In *Demokratsija*, January 2001.

Shishmanov, Dimităr. *Neobiknovenata istorija na maloazijskite bylgari* [The Extraordinary Story of Bulgatians in Asia Minor]. Sofia: Poni, 2000.

Tsoncheva, Mara. *Tsărkvata "Sv. Georgi" v Sofija* [The Church of St. George in Sofia]. Sofia: Bălgarski Khudozhnik, 1979.

City Perception in Transit: The Prague Experience

Joanna Derdowska

Abstract

This article focuses on a poetics of re-enchanting the city which perceives the apparition of new urban mythologies and the process of recreating new enchantment of the urban world in its Weberian meaning. The article summarizes a socio-political basis of the notion of *magic Prague* and provides the main positions of its critique. Secondly, the new utilisation of the emblematic figures of the Prague myth by the tourist industry is shown. Finally, the surrealists' city games conducted in Prague are interpreted. The main lines of the analysis concern two reading levels: first, presenting a narrative process as a walk and distinguishing a blasé walk of modernity from a present-day active walk as defined by Michel de Certeau. Second, interest is taken in the "anatomy of the city" represented by a figure of a map as a way of providing structure and as a milieu of a game. The signification of a city's spaces often draws upon its myths and legends, which are historically redefined and recreated. Great changes are undertaken in this area during periods of economical and cultural reorganisation. The development of a few Prague myths will serve as an example of these processes.

The most considerable myth of the Czech capital city is the one of "the magic Prague." It has been widely propagated by an Italian bohemicist Mario Angelo Ripellino in his book entitled *The Magic Prague*, first published in Italy in 1973. Ripellino describes the alchemical, mysterious face of the city, where he himself becomes a scribe, a kind of a writer of memoirs, who saves this fading image of the mysterious city disappearing under the pressure of industrial civilization and the greyness of the communist reality of the 1970s. Ripellino recalls the legends of the Jewish Golem and the alchemists; he portrays the writers and the literary characters from their books (Hašek and Good Soldier Švejk, Kafka and Joseph K.) as if they were all a part of the same half-fictive world. However, the book can be easily criticized for being single-tracked and even kitsch as it straddles the border of documentary and lyric fiction and gives rather an emotional image of Prague. A few critics accuse Ripellino of creating a cultural myth of the city that would not correspond with the real state. Besides, Ripellino is reproached for his language, which is so full of pathos that in today's point of view it seems very distant indeed and at some places it really becomes more Czech-centred and wrapped up than many Czech writings of the nineteenth century. Ripellino was, however, describing the Prague of the 1960s and 1970s, which was rather a phantom since it was impossible for him to visit the communist city. Thus, even for him the reality was slowly ceasing under the pressure of its own myth (and it was not only Ripellino who created visions of this sort; Prague as a glorious city is also commended by Christian Norberg Schulz, an architect phenomenologist who focuses his interest on the "mysterious" aspects of its architecture). An interesting polemics with the idea of "magic Prague" consists in showing its socio-psychological motivation of glorifying such a closed and far away city as a kind of a positive ghetto and describing it as a sentimental and nostalgic Golden Land. Such a critique has been presented among others by Peter Demetz, a Czech historian who emigrated to the U.S. According to Demetz, the old formula that used to give a drop of hope during the dull and intellectually unfertile period of the so-called normalisation, is now useless and becomes an ordinary product for sale. Let's consider if these presumptions could be right.

After 1989 Prague became a popular tourist destination, not expensive from the Western point of view and in some senses exotic. How attractive it is can be indicated by the size of the American minority that has emerged there for the last 20 years. Tourists often take with them Ripellino's book, as a guide to reveal the supposed mysteries of the city. A subtitle in the French translation of the book describes it even as an "initiatory guide to Prague."[1] But the adepts of such an initiation must be rather disappointed.

One of the Prague icons is Franz Kafka, a talented Jew writing in German and living in Prague, whose literature was actually discovered and offered to wider public only long after his death, which also contributed to

[1] Ripellino is not the only author who creates a vision of this sort. Prague as a glorious city is also commended by Christian Norberg-Schultz, an architect-phenomenologist who focuses his interest on the "mysterious" aspects of the Prague architecture (Norberg-Schulz).

his own myth in the Prague of the 1970s to the extent that in Ripellino's Prague Kafka was one of the essential figures. In today's city, one can feel that the writer's character has become one of the personages in the freakshow of Prague's legendary heroes. A tourist agency spreads a leaflet advertising "Prague of legends, Kafka and mystery," which makes the writer's name a synonym for the mystery (similarly as the word "Kafkaesque" describes strange feelings in the common language). Another says that a city tour it offers is a "selection of the most interesting and enigmatic stories – partly fictive, partly historic – which are still the centre of the *popular imagination.*" Here, the artificial character of this created world is even honestly confessed, especially in the notion of the "popular imagination," which evokes all the ideas of pop culture and longed-for sensation tabloids (images 1 and 2). Needless to say, Kafka's portraits are present on boxes of matches, T-shirts and ballpoint pens. A sun-tanned image of Kafka (who actually died of tuberculosis) was even a symbol of the Prague Marathon in 2003. To paraphrase the definition of social realism, which was socialist in form and local in content – the tourist images of many well-known and visited cities are indeed Disneyesque in form and local in content. Although I am very far from condemning the tourist industry for falsifying someone's biography, my point is that in this mysterious (and mystifying) poetics of city-storytelling, Kafka is not only mistaken for his literary alter egos, but is slightly confused with other characters of the Prague myth-cocktail such as the Prague Golem or Švejk.

There commences a debate between the city government, which appreciates the touristic movement as much as any other business and the inhabitants, who, generally speaking, do not like tourists for various reasons. Once more I underline that it would be useless to simply berate all that is new in the city and that does not match its sentimental image that very often has also been created artificially. During a conference devoted to Prague's architecture at the beginning of the 1990s, Jacques Derrida was absolutely right in claiming that no vivid city could serve as a lifeless gallery or sanctuary. One should, though, consider in what case a city becomes such a sanctuary: is it when it produces its exaggerated images for the tourist industry as in the case of the just presented Prague icons; or maybe it would rest even more paralysed if it were to stay intact and serve as an open-air folk museum of the older, so-called authentic myths? I tend to opt for the idea that tourism, paradoxically, turns out to preserve some fairylike phantoms, no matter how banal and trivial they become.

Nevertheless, signs of some innovative searches for the city's magic can be seen in some literary and artistic productions that seem to create and rediscover the attraction of Prague by its re-enchantment.[2] One of the ways of reviving the perception of a city are interpretative games of a different kind. The tradition of discovering and exploring the city space in the quest of its mysteries dates at least to the French surrealists' movement and the narratives of André Breton or Louis Aragon[3] that found the successors in the psychogeography created by the situationits in the 1960s (Sadler) and, on the other hand, in writings by Georges Perec. An example of contemporary Prague city-space games would be a project entitled *Ludus Pragensis*, directed by authors connected with the Prague surrealists group. The director of the game has set in the city plan a figure determined by the settlements of the participants and has chosen three objects inside it; he visited these places, took the photos of them and gave an "*imaginary interpretation* attempting to handle its genius loci" (Purš 100). The other participants were to find the places that they thought could match the description and to give their own sketches of them. The ambiance of the game can be evinced by the following quotation describing the Jubilee Synagogue (image 3):

> It is a dusty cake with whipped cream, a dessert in orient style, a wired mixture of the sacred, pomposity and decline. A syncretic cramp from the era when the new was mixing up with the old, the living with the dead. Digestible tiered stories, lux in Oriente in the shadow of begrimed windows, the templars producing bonbons with an ingenious sales-name 'Salomon's wisdom' (Purš 102).

According to the author, the game's aim was to verify whether the city could function as a "generator of analogical thinking." Taking part in such a game forces the participants to search, to observe and to interpret. In a way it poeticizes the city:

> [It is] the way through the city was actually nothing but a glimpse into the mirror. Maybe the city is a mirror [...] Let's prevent Prague from being a comfortable den, because it never was and never will be its purpose. You

---

[2] I develop the term from Max Weber who was talking about two disenchantments of the world: the first after the fall of the syncretic magical praxis, the other after the coming of the rational era. The new enchantment are the contemporary mystical movements and religions.

[3] A. Breton's *Tournesol*, *Nadja* or *Amour fou*. L. Aragon gives a description for this kind of city perception: "metaphysique des lieux" ("the metaphysics of places") that shows the places in the imaginary poetics (Decottignies).

must feel the ache in the legs and the head because of the Prague streets – otherwise it will grow together with everydayness (Purš 106).[4]

A city walk, derived from nineteenth century flânerie, is often perceived as a means of creating sense. For a French theorist Michel de Certeau, for example, a walk has its narrative dimensions (in example when taking my way to see a play of a Russian writer about Stalinist concentration camps, that took place at Lenin Street, I had to change at Stalingrad metro-station – here we have a little piece of soviet pedestrian narrative). In his text *Walking in the city*, de Certeau interprets the itineraries of pedestrians as a piece of information or even as a piece of art. He assumes that such strolls can be undertaken by chance (as my way to the theatre) as well as even without the consciousness of the walker. The Prague surrealists, similar to their French predecessors, plan their walks intentionally but at the same time they let the space guide them: "A surrealist poet doesn't stroll like a repulsive Sunday bourgeois, he is *strolled*, by a mysterious force to which he intrusts himself to be at a disposal of a surprise, a meeting or a miracle" (Clébert 433). "My steps carry me," as Breton says. The same can be said about the contemporary city games. Thus, also the way of treating the city plan, its material graphic representation, is very different than that of a common use. Firstly, the space in the map is individualised, in our example it is given by the residence of the participants. The city space ceases to be an objective net of points and the user does not search the shortest possible way somewhere but instead he sets off to look for a "pleasure of the space" (in analogy to Roland Barthes's "pleasure of the text"), the space that does not serve to commute but provides an opportunity to be explored. In *Ludus Pragensis* the map is interpreted (and even defined) by the inhabitants and players, with no attempt at objectivity.

The "Surrealistic places are places that are found," (Clébert) as in Martin Stejskal's project of the Prague chronicle of the end of the millennium, constructed by taking photographs of a Prague sewer cap over 12 months (image 4). The outcome was a series of images that demonstrate the influences of weather, human beings or casual conditions on the appearance of this piece of pavement, and were loosely interpreted as predictions for the city. The searchers do not look for something marvelous but rather for something banal that can be found everyday in a

---

[4] Another game described in detail in the same issue consists in visiting and exploring a place in the city that none of the participants knows (an island on Vltava River in the centre of Prague) and comparing it with a previous image.

noisy street. "Surrealists de-realize the city ceaselessly" (Blancquart 29) and that is the exact sense of its re-enchanting.

Even if such actions oscillate between children's games and the mentality of a so-called backpacker or a tourist who intentionally avoids the places recommended in tourist guides and they may to some extent appear banal, their reconnaissance of the urban areas that they offer are indeed poetical, full of sensibility and fantasy. The city space, instead of being a neutral net of places, becomes an object (if not a subject) of a game and its location.

As a last remark, it should be also mentioned here that the continuity of the re-enchantment is also maintained in the literary production, and not only in the one affiliated wit the surrealist movement. For example, the novels of Michal Ajvaz, who stylizes the city space in a Borgesian way, let it have many dimensions and guide the movement of the characters:

> Is it possible that we live in close proximity to a world teeming with strange life, a world which perhaps has been here since before our city existed and yet which we know nothing about? The more I have thought about it, the more I have been inclined to admit that it is very well possible, that it goes with our way of life, with how we live within demarcated circles we're afraid to leave. (Ajvaz)

To conclude, the notion of "magical Prague" cannot be simply erased from the character of Prague. Nevertheless, it should be considered that it evokes a much wider spectrum of problematical meanings than it would seem. Simply put: the primary "magic" image worked up mainly in the second half of the twentieth century has now been degraded and banalized by being used in a mass context and by commercialisation. However, in many circles, particularly among artists, various initiatives are being taken to stimulate the perception of the aspects of the city space connected with its extraordinariness. The sensitive habitants (and the sensitive guests as well) can hardly accept the vision of loosing the allure and the savour of mystery. The efforts to save it are being held mostly in the micro-scale, on one's own account or for a narrow group of friends, which also corresponds with the dominance of individual, authentic feeling, which connects the city-player with the model of the modernistic pilgrim.

Works Cited

Ajvaz, Michal. *The Other City / Druhé město.* Extract (Tatiana Firkušný and Veronique Firkušný-Callegari, transl.), thanks to Dana Blatná Literary Agency: http://www.dbagency.cz/index.php?s=book&prid=64&a=extract&name=michal-ajvaz--the-other-city-/%A0druhe-mesto

Aragon, Louis. *Le paysan de Paris*, Paris: Gallimard. 1926.

Blancquart, Marie-Claire. *Paris des surréalistes.* Paris: Snela. Édition de la Différence, 2004.

Certeau, Michel de. *L'invention du cotidien. Vol I. Arts de faire.* Nouvelle édition, établie et présentée par Luce Giard. Paris: Gallimard, 1990.

Breton André. *Nadja.* Paris: Gallimard, 1972.

—. *L'Amour fou.* Paris: Gallimard, 2001.

—. *Clair de terre.* Paris: Gallimard, 1923.

Clébert, Jean-Paul. *Dictionnaire du surréalisme.* Paris: Seuil, 1994.

Decottignies, Jean. *L'invention de la poesie – Breton, Aragon, Duchamp.* Lille: Presses universitaires de Lille, 1994.

Demetz, Peter. *Prague in Black and Gold: The History of a City: Scenes from the Life of a European City.* New York: Hill & Wang, 1997.

Derrida, Jacques. "Generace (jednoho) města. Paměť, proroctví, odpovědňosti." Transl. Karel Thein. *Tvar. Literární týdeník* 16 (1991): 12-13.

Norberg-Schulz, Christian. *Genius Loci. Towards a Phenomenology of Architecture.* New York: Rizzoli, 1980.

Perec, Georges. *Espèces d'espaces.* Paris: Denoël-Gonthier, 1976.

Purš, Ivo. "Ludus pragensis. Interpretační hra." *Analogon* 18 (1996).

Ripellino, Angelo Maria. *Magic Prague.* Berkeley: University of California Press, 1993.

Sadler, Simon. *The Situationist City.* London: The MIT Press, 1998.

Stejskal, Martin. "Kronika z konce tisíciletí." *Cesty za hvězdou. Zrcadlo hermetismu v české krajině.* Praha-Litomyšl: Paseka, 2001.

Images

Figure 1: Franz Kafka at the Jalta hotel billboard

Figure 2: Kafka as a city icon.

Figure 3: The Jubilee Synagogue.

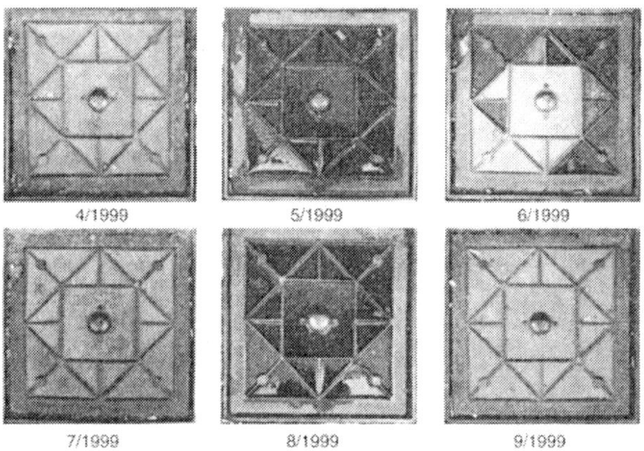

Figure 4: Martin Stejskal's Chronicle of the end of millenniu. This images comes from the book by M. Stejskal cited in the text.

THE PSYCHOANALYTIC TOURIST: DISPLACEMENT AND PLACE OF THE OTHER

GÜL BAHAR CÖMERT-AGOURIDAS
AND DIMITRIS AGOURIDAS

Abstract

We wonder whether it is not tourism that makes the positioning of the signifier "tourist" in this enunciation possible, but rather the signifier emerging in that position that makes possible the act of tourism. Since the subject exists as ex-istant, the dialectics of distance — difference echoed in space — and of being-out-of-place will define the horizon of subjectivity. In tourism, distance is experienced as traumatically as the alienation of narcissistic identification. We raise the question of whether it could not be precisely the reassertion of an initial position that is the desire of the tourist. The resonance to the "*Wo es war, soll ich werden,*" is not accidental, since the *Wo*, the "where," indicates the possibility of processes through which the symbolic construction of space lends itself to uses that support psychoanalytic discourse, and opens up a field where tourism and the subject can be articulated.

It would be useless to deny that there is a kind of striking categorical *un*-similarity between tourism and psychoanalysis, making any articulation of the two, if this would be the required task, a quite problematic one.

Perhaps this is because the problem of articulation continues being posed, to a great extent, with a somewhat exaggerated exclusiveness, as a problem of transference or exchange, in the best of occasions, between fields of knowledge, and, indeed, often without problematising the subject of knowledge, and what the desire of articulation, or even articulation itself as desire, has to reveal about that subject. It would be, however, possible to form a discourse on articulation within a different frame, especially if this would allow us to discern behind the philosophical and scientific object the emergence of an object of desire; and this in a manner which could very well lead us to the questioning of objectivity as the defining principle of our relation to reality. And, to follow Milner, in this fashion, instead of obtaining its ideal *from* science, psychoanalysis would rather create an analytical ideal *for* science, one of discourse, by necessity.

In this fashion, only when tourism as a concept is cleansed of its established "industrial" references, only through its de-anchoring from commercially imposed optics (in times that merchants and other forces of the market feel compelled to teach us what the meaning of knowledge is), and when it is expanded to such a degree that it finally means nothing, only then will it be able to mean something for the subject. Only when the motion of nostalgia, and the nostalgia of all motion, with whatever nostalgia allows to emerge, this άλγος of νόστος, the pain of return to a forever lost home, subdues the study of socioeconomic impacts of the temporary corporeal displacement within some industrial or postindustrial structure, and breaks with any vocational bonds, we may be able to learn something about the subject from tourism. Only when tourism as such is posed as a problem and a request for truth, and its existence is denied, only then will its subject appear.

Hence, it is psychoanalysis that is chosen as the privileged analytical instrument for an understanding of the desire at the base of all tourism, traversing it all along, and subjecting any attempt to understand tourism to its analytical principle.

In the orientation of our investigation towards that which is displaced onto the temporary spatial, corporeal displacement that tourism appears to be, one can, without much difficulty, recognise something of the psychoanalytical discourse organising our gaze upon it. The displacement of the body takes here the place of the body itself; if something is displaced upon the body in the case of hysteria, for example, there is some sort of legitimacy in an attempt to look for something displaced upon the displacement of the body, in the case of tourism. Because, what proceeds through the opposition between Τόπος (the place of corporeality) and Λόγος (the word, sense, the symbolic), results in a topology, the practice of which is the treatment of representation with the body, signifying the inscription of this very practice within the totality of our fantasmatic productions (Nasio).

It is in, by and through the discourse of displacement that (touristic) desire will be structured, and the act of placing that desire at the centre of the analysis of tourism derives from the study of psychoanalysis, to the extent that the latter has established the discourse of displacement as its central tautology, since if analysis has taught us something, this is that discourse is displacement.

The fact that tourism has remained virtually unexplored through such a prism, despite the very few interesting, yet sporadic relevant attempts (eg. Haddad & Haddad, 1995; MacCannell, 2001; Kingsbury, 2005), seems to be quite paradoxical, given the ever growing importance that tourism, qua discourse of displacement, seems to be enjoying in modernity, to such a degree that McCannell (1999) will be able to discern in the tourist the model of the modern subject. It may not be coincidental that at precisely the same period hysteria would become (re)cognisable in the context of that very discourse of displacement, of *Verschiebung*.

If McCannell is right, and, despite the objections his thesis has raised (Kaplan), the tourist is, indeed, the model of the modern subject, nostalgic of itself as he portrays it, and if, as Lacan (1966) has, in every possible tone, asserted, the subject upon which psychoanalysis is exercised is exactly the modern, Cartesian subject, the subject of science, it follows that the tourist is also the model of the subject upon which analysis is exercised. If when we speak of a subject of tourism we most frequently have no idea what we're talking about, it is most probably due to the fact that we forget that it is the Cartesian subject, the subject of the *cogito*, that we are concerned with.

However, our analysis of that subject cannot be confined to the "discovery" of a latent content in tourism out of the reach of the tourist's consciousness, to a speculation on the meaning of tourism, which would almost definitely regress from a potential meta-psychological study of tourism to the construction of yet another tourist-motivation model.

Even with every single meaning of such an action unveiled, the action itself remains an enigma, simply because the labour to be ultimately undertaken is not one of translation – translating the latent into the apparent. If there is something to be articulated that would be the question of the form itself, always imprinted in the seemingly simple phrase "I need a vacation." This, of course, doesn't really refer to any need – any claim concerning the existence of a "tourist need" wouldn't fall short of outrageous, an outrageous biologisation of tourism - but is already a demand, for which no object that would satisfy it can be located. There is no object "vacation," as anything else than what it etymologically implies.

If, nevertheless, we can conceive the Cartesian subject as one thriving in a nostalgia for itself – and it remains to be seen if indeed we can – and if the same modern subject is the one that can enunciate "I feel a Tourist in my Life," we are confronted with a necessary dialectic development: We have only to wonder and examine whether it is not the tourist, in the act of tourism on the level of reality, that makes the positioning of the signifier "tourist" in this enunciation possible, but rather the signifier emerging in that position that makes possible the act of tourism. In this way, touristic subjectivity becomes not only a reference to the subject in the act or performance of tourism, but also a digressive possibility that the modern subject is per se touristic, which has to be explored.

The concept of distance needs to be investigated in this respect, since the specular relation distances the subject from itself, although a rapprochement of the Ego with the specular image by no means offers a solution to the problems emerging from this distance (Safouan), and this because the subject itself emerges *in* distance. Since the subject exists as ex-sistant, the dialectics of distance – the name of difference echoed in space – and of being-out-of-place will define the horizon of subjectivity.

In his seminar of 1966-1967, entitled "The Logic of Fantasy," Lacan, drawing legitimacy from the psychoanalytic discovery of a "knowledge which doesn't know itself" (Lacan 1975, 122), will perform a memorable manipulation of the cogito through DeMorgan's rules to demonstrate that "either I don't think, or I am not," later expanding it to conclude that "I am where I don't think, I think where I am not."

Material, corporeal displacement will result in a status of a temporary being-out-of-place, equally material to begin with, addressing, however, the question of the subject's consistency of being-in-the-world, further displacing the latter onto the dialectics of space, in what could probably be an attempt to put a limit on it. One has to keep in mind, of course, that the dialectics of space in operation are not merely expressed by the opposition of here to there, but also by that of internal to external, and that, in the latter case, the external is not given as a projection of the subject's internal, of the subject's drives (Safouan), but rather "as the position or the place where the desire of the Other is positioned, there where the subject will meet it" (Lacan 1988, 272).

"What is the Other?" we may then ask, along with Lacan. "It is this field of the truth that I have defined as being the place where the discourse of the subject would gain its consistence, and where it stands in an offer to be refused or not" (Lacan 2006, 24).

The question that can be raised in relation to this is whether it could not be precisely the reassertion of the initial position that orients the desire of the tourist. I leave only to return to where I already was, to re-occupy the place that I already occupied, to affirm my identification to my proper place. The resonance to the *Wo es war, soll ich werden*, where It was must I be, is not accidental, since the *Wo*, the "where" of the dictum, indicates the possibility of processes through which the symbolic construction of space lends itself to metaphoric uses that have supported the psychoanalytic discourse of displacement.

Parallels can also be drawn between this reading of touristic desire to the *fort-da*, another metaphor illustrating, if nothing else, how the spatial becomes a symbolic construction: I let go, objectify myself and throw it away, in order to re-appropriate it, to pull it back. I try to recover a unity, to de-alienate myself from my body by displacement. Something hides in order to be found, gets thrown away in order to be pulled back.

In a nutshell, I become a tourist, because I will, by necessity, also learn that I have to become a re-tourist. Despite my claims about the reasons of my trip, which will most probably be accurate as regards the subject of the sentence in which they take place, it is the production of this knowledge, on the level of the desiring subject, the subject of the enunciation, which manifests itself as Reason, in revealing that mastery over the body – by the very act of returning – constitutes a renouncement (Kaufmann).

Had I not returned, had I gone on with the pleasure of being there, this pleasure would turn into a suffocating, painful thing of death, depriving me of any possibility to desire being there; therefore, I return, I cut it off myself, and give myself the chance to have a "there" that is not "here," offering an opportunity to the spatial (in this case) dialectics of desire to function. It is practically the structural necessity of castration that the produced knowledge reveals: I, as the master of my body, have to be castrated in order to be able to desire and have the ontological status of a desiring subject. In this manner, though, the subject also discloses the failure of mastery over the body - that is castration - which, paradoxically, is also the condition of the existence of any mastery.

This is one way to read what Freud has described as "A Disturbance of Memory on the Acropolis," one of the earliest accounts of what has later become known as the Stendhal Syndrome (Magherini): it is the threat of a pleasure too intense to be pleasurable that inhibits him at Trieste, where he is confronted with the suggestion to proceed to Athens, of a pleasure that will have been no pleasure any longer, because it has crossed a limit that is set by a master constituted in paternal impotence. Freud's being there functions as an interrogation that will bring to light the impossibility of mastery, the fact that the master is, indeed, castrated.

The Acropolis is an unattainable thing of desire, the meeting with which can never take place, an ever-missed object. It can only be a limit which can never be reached (Žižek). It was as early as the time of Seneca that this knowledge became available, when he attested that man travels in search of an eluding object (Urry). This is what ascribes, in the mind of the young student Freud, the character of the sublime to Acropolis. It is a sublime object that exists only as elusive, the existence of which is guaranteed by its unattainability, the latter guaranteeing, in its turn, the possibility of desire. The object Acropolis seems to be a maternal thing, and as such not only unattainable, but also forbidden.

The knowledge produced by the travel to Athens, and the visit to the Acropolis, expressed by the realisation that it, the Acropolis, really does exist, is the knowledge that the master, that is represented by the father in this case, could have never attained it, could have never been the desire of the mother. Hence the visit to the Acropolis becomes an interrogation of the master, regarding the structural necessity of the master's castration.

In a bizarre way, this disturbance of memory on the Acropolis reflects the problematics of the question itself of the subject of the cogito: it is one thing to think of the Acropolis, and another to be there. This is yet another way given to us to approach the manner in which the discourse of displacement gains its existence within the entanglement of what is spatial with what is of the subject. Indeed, what is seen here is the splitting of the subject, finding a support to symbolically emerge, in the presence of what can only be inscribed as absence and cannot be symbolised. The Acropolis keeps the role of that impossible object. It stands there not as an object of reality, not as a symbolic object, but as a real object, that managed to become inscribed only because there was a failure in the process of formalisation, an object that, had it not been for the question Freud had to pose by his travel, and the failure that travel would evoke, would remain one that doesn't cease not to be inscribed (Lacan, 1975).

In the last resort, it was the enjoyment mediating on the knowledge Freud the tourist gained, that the master, in the face of his father, is castrated, that was the non-sense of his trip, filled with guilt and pain, as all such enjoyment is.

But, to return to vacation and the void of tourism, to which vacation is a reference, this is not a void left behind, or even one to be carried along in the travel, but rather the effect created by a one-boundary surface appearing to have two instead. Tourism, the name we have given displacement, is an act of speech, the violence of a cut along the centre of a Möbius strip[1], turning it to a strip with two boundaries, cancelling its fundamental properties, so that the signifier represents the subject while, at the same time, it makes it disappear (Nasio).

I am condemned to be the representative of myself in my word, by means of the "me" that I am obliged to use, which is an imaginary construction by which the subject objectifies itself through its own

[1] A one-sided non-orientable surface obtained by cutting a closed band into a single strip, reattaching the two ends with a half twist.

representatives (Dor, 1985, 1992). Indeed, as Lacan has it, "The misfortune of the subject in word ['the drama,' he calls it] is that it proves there its lack-in-being" (Lacan 1966, 655), given the relation that the subject forms with itself through and by the signifying order, a word in which the "me" mistakes itself for "I."

A true Other is necessarily supposed, an Other like the subject itself. And, in the same manner that the subject's symbolic registration suggests that the subject will be only represented in its word by a signifier of itself, and, therefore, the true subject will be forever eluding words, the Other, too, will be present there through a representative, a signifier. Furthermore, inasmuch as the subject is only represented in word, the signifier is nothing but the "symbol of an absence" (Lacan 1966, 24). Wherever the signifier is, the subject is not, with the substantial consequence of the subject and the Other becoming positioned on the two opposite sides of a wall called "Language" (Lacan 1978), the order of which presupposes the absence of the true subject. It is not by accident, so to speak, that the word "sign" in ancient Greek (Σήμα) means "sign" only as much as it also means "tomb," "grave." To name is to name an absence, and to signify is to signify death. This relation between signification and death justifies the accreditation of the character of *joui-sens* to the signifying chain (Lacan, 2001), as well as the situation of the signifier at the level of the *substance jouissante* (Lacan, 1975).

However, something always eludes the other and within the Other. Therefore, although the signifier hides the truth of the subject, it is, at the same time, supported by it. Nevertheless, it is the signifier that seems to be speaking the truth, to be conveying meaning to the other, to be the subject that acts, although it is but a mere agent, one, moreover, that has no idea about the true meaning of the message it conveys.

Desire - constitutively alienated in the desire of the other, for the sole reason that its first object is to be recognised by the other (Lacan, 1966), to be the desire of the other, to be the other's phallus, the castration of which gives birth to the subject as such, and which, thus, bars the subject – is produced in the *au-delà* of demand. Because the demand initiating the discourse, driving it, is the communication of two subjects, an impossible task, echoing, if not being identified with, the demand to be the Other's desire. What remains of it is the object a, and it is the object a, as object cause of desire, that is produced by the discourse.

In a sense then, tourism is nothing more and nothing less than a reversed symptom, than a symptom turned inside out, not the opposite of hysteria, but rather some kind of counter-hysteria, starting from the end to make a circle all the way to the beginning, that is hysteria par excellence.

The hysteric is a tourist of her own life, in her own right, in the sense that her body becomes a produce-able and consumable landscape of an unaccountable picturesque. The difference between the tourist and the hysteric is the difference between hospitality and the hospital.

But, still, within the discourse of the tourist, the tourist is a mere shadow of the gaze. And the tourist gaze is not seeing; it is, on the contrary, showing. It is not the manner in which the tourist appropriates the landscape, the picturesque, let alone the picture. It is, reversely, the manner in which the tourist is appropriated by and objectified within the landscape. The tourist gaze is not understood when I gaze myself in the mirror. I will confront it when, after the introduction of one more mirror, I will see myself being gazed at. It is the shadow of the gaze, confronted like that, as an object to be gazed upon, that the tourist is. The gaze is what introduces the tourist into the picturesque, regardless of the form the latter may take, something ever outside the picture dropping its shadow – namely the tourist – inside the picture. If tourism is unfinished, or unfinalised, this is because the gaze cannot return to the subject, the subject will not recover the gaze. As Lacan puts it, "We can capture this privilege of the gaze in the function of desire, rolling, if I might say, along the veins from where the domain of vision has been integrated within the field of desire" (Lacan 1973, 98).

The critical dimension is that of the relation of tourism to the *fantasme*, formulated as $◊a (Lacan 1966). Tourism expresses the latter, given that the supposition of an Other in tourism is the supposition of an incomplete Other, necessary for this imaginary effect to take place. As fantasmatic, tourism seeks to invade the Other, like performance would invade a scene, to fill out a void in the Other, which, although not really there, is there in the Real. The picture of the picturesque is the picture including the tourist himself, by means of the tourist's absence. It functions as Other as long as it includes an offer-able void to be filled in by the touristic *fantasme*.

Whatever the tourist may be trying to retrieve, to re-cover, he will always be coming face to face with an always-already covered object. Whatever the tourist gazes at is the gaze of whatever gazes back at the tourist, the thing of which the tourist is the shadow. Therefore, the tourist will always re-turn, he will always be a re-tourist. However, the re-tourist will always re-turn as a symptom. Externality, exclusion from the truth of the - hence lacking - Other, the truth of whom is that he does not possess my truth, introduces the subject as an integral part of the Other's game. My externality is internal to the Other (Žižek). The *objet petit a*, the tourist gaze, is this secret of the Other, eluding the Other as much as within the Other.

For the tourist, distance is experienced as traumatically as the alienation of narcissistic identification. In fact, distance is part of the process. And it is experienced as a difference, the geographic difference between the here and the not-here, the here and the elsewhere, the place of the Other, metonymic of the linguistic difference between the subject and its representation. The question of the tourist, articulated through the discursive device, is taking up the form – "Where is (it) ... missing?" in which the "where" is the indicating instance of the metonymy of desire, supported by the "missing" of the subject that the signifying question "where is (it)" will be hiding, and only as long as the persistence of the "missing" will be kept concealed can tourism be articulated.

The fundamental presupposition of desire is absence. There must be something missing which will be desired, something eluding to be run after. This is the object cause of desire. But this object is positioned as such by desire itself retrospectively, to the extent that desire itself is its own object, identified with its interpretation. Distance, the spatial distance to be covered, to bring the body of the tourist to the place of the Other, functions as nothing but the metonymy of difference, and, in being so manifestly metaphoric, this *metaphora* of the tourist body is what characterises tourism.

The "here" positioned at the place of the Other, signifies the tourist's externality, his exclusion from the truth of the Other. The "here" must be elsewhere, since what is at stake seems to be the "where." The "it is" is the object cause of desire, the object gaze in the tourist's discourse, in disjunction with the missing of the subject. Hence, in trying to capture desire, the tourist will constantly end up being captured by it. This is why instead of photographing, I am being photographed. Because the tourist gaze is the desire to gaze, and the tourist desire is the gaze itself as desire - the gaze gazing at itself.

Works Cited

Dor, J. *Introduction à la Lecture de Lacan, tome 1: L'Inconscient Structuré comme un Langage*. Paris: Denoël, 1985.

—. *Introduction à la Lecture de Lacan, tome 2: La Structure du Sujet*. Paris: Denoël, 1992.

Haddad, A. & Haddad, G. *Freud en Italie: Psychanalyse du Voyage*. Paris: Editions Albin Michel, 1995.

Kaplan, C. *Questions of Travel : Postmodern Discourses of Displacement*. Durham, North Carolina: Duke University Press, 1996.

Kaufmann, P. ed. *L'Apport Freudien : Éléments pour une Encyclopédie de la Psychanalyse*. Paris: Bordas, 1993.

Kingsbury, P. "Jamaican Tourism and the Politics of Enjoyment". Geoforum, 36 (2005): 113-132.

Lacan, J. *Autres* Ecrits. Paris: Seuil, 2001.

—. *Écrits*. Paris: Seuil, 1966.

—. *Le Séminaire – Livre II : Le Moi dans la Théorie de Freud et dans la Technique de la Psychanalyse*. Paris: Seuil, 1978.

—. *Le Séminaire – Livre V: Les Formations de l'Inconscient*. Paris: Seuil, 1988.

—. *Le Séminaire – Livre XI : Les Quatre Concepts Fondamentaux de la Psychanalyse*. Paris: Seuil, 1973.

—. *Le Séminaire – Livre XVI: D'un Autre à l'autre*. Paris: Seuil, 2006.

—. *Le Séminaire – Livre XX : Encore*. Paris: Seuil, 1975.

MacCannell, D. *The Tourist: A New Theory of the Leisure Class*. Berkeley: University of California Press, 1999.

—. "Tourist Agency." *Tourist Studies*, 1(1) (2001): 23-37.

Magherini, G. *La sindrome di Stendhal: Il Malessere del Viaggiatore di Fronte alla Grandezza dell'Arte*. Milano: Ponte alle Grazie, 2003.

Milner, J. C. *L'Œuvre Claire: Lacan, la Science, la Philosophie*. Paris: Seuil, 1995.

Nasio, J. D. *Les Yeux de Laure: Transfert, Objet 'a' et Topologie dans la Théorie de J. Lacan*. Paris: Champ-Flammarion, 1995.

Ross, G. F. 2nd ed. *The Psychology of Tourism*. Elsternwick: Hospitality Press, 1998.

Safouan, M. *Lacaniana – Les Séminaires de Jacques Lacan, tome 1: 1953 – 1963*. Paris: Fayard, 2001.

Urry, J. *The Tourist Gaze: Leisure and Travel in Contemporary Societies*. London: Sage, 1990.

Žižek, S. *The Sublime Object of Ideology*. London: Verso, 1989.

RECONSTRUCTIONS OF PLACE: OSMAN HAMDI, OTTOMAN IMPERIAL VISUAL CULTURE AND THE TERRITORIALIZATION OF THE GAZE

LEWIS JOHNSON

Abstract

This paper situates some of the archaeological, photographic and artistic work of the Ottoman Osman Hamdi as variously marked by a late nineteenth century inter-imperial politics of place. Using the work of Maurice Blanchot on the image as indicator of a problematic of place, this paper analyzes the performative enactments of Osman Hamdi, as staged in photographs of him at excavations at Mount Nemrud in 1883, of dilemmas concerning archaeological remains and place. This is contrasted with photographs of western archaeologists heroically delimiting and restoring place as archaeological site, and photographs of archaeological remains as if salvaged from a decline into materiality. This is further contrasted with an Ottoman politics of place, as suggested in Hamdi's earlier 1873 photographic project of Ottoman dress, and related to the mode of articulation of the display of archaeological remains from across the empire in the Ottoman Imperial Museum from 1891 onwards.

In the second of three appendices to *L'espace littéraire*, translated into English as "The Two Versions of the Imaginary", Maurice Blanchot pursues a comparison between the image and the corpse the terms of which provide some indications concerning the axioms that I shall be using to supply a framework for a consideration of the evidentialities of images in relation to questions of place. This comparison offers a way of understanding a certain placelessness of the effect of images–and, perhaps, given the context and terms of Blanchot's argument in that appendix, of literary imagery too–that can, I believe, serve an understanding both of the dynamics of spatial displacements and the disturbances of place that are associated with discourses of modernity and of ways in which senses of place can be reconstructed. Open to radical loss, senses of place can serve to release blockages in thinking and acting associated with debates about globalization and regionalism, at least in fields of communication and culture.

My topic involves me in a number of ways in the "reconstructions of place" of my title. I want to show that certain photographs of Osman Hamdi, taken at Mount Nemrud and in Sidon, provide us with the possibility of an experience of place and placelessness that traverse the demands of the modern discipline of archaeology, if not of history, as well as the then pressing demands of inter-imperial cultural politics. I shall be reconstructing something of the reconstructions of place that Hamdi served, both at the site of the "discovery" in 1883 of the massive heads of, among others, Zeus-Oromasdes and Apollo-Mithras, Greco-Persian gods of Antiochus I, King of Commagene, in the reassembling of the remains of what is called the hierothesion, or "dwelling place of the gods," constructed some time during Antiochus' reign from 62 to 32 BC, and, later, in Sidon as well as subsequently in the construction of what is now the main building of the Istanbul Archaeological Museum, finished and opened in 1891, to house the most renowned of Ottoman archaeological finds, the so-called *Alexander Sarcophagus*.

The experience of place and placelessness that I hope to isolate, as possibility, in the photographs taken of Osman Hamdi as on-site archaeologist will, if I am right, have a bearing on how we understand the display of the remains here, in Istanbul at the Archaeological Museum. The possibility of such an experience also has a bearing on an understanding of Hamdi's work as a painter, in particular in his repeated painting of figures resembling himself in his paintings [fig. 1], usually, I understand, deriving from photographs of him, wearing traditional Ottoman dress rather than his more usual jacket, waistcoat, shirt and tie and trousers of Western European inspired cosmopolitan styles [fig. 2]. Such paintings, along with the photography of archaeological sites and the construction of museums and arrangement of their collections, may also be treated as participating in reconstructions and deconstructions of place, I think; but the strictures suggested by Blanchot would need to be followed. So, I shall try to establish them as succinctly as I can.

In treating Hamdi's paintings, that resemble or resembled him, and, by extension, the variously active Hamdi himself, we might follow Blanchot's comparison of image and corpse in its more difficult passages. As, for example, when he, Blanchot, stipulates that the living man does not

Figure 1 Osman Hamdi, *Ab-ı Hayat Çeşmesi* [*Fountain of the Water of Life*], oil on canvas, 1904 [reproduction on display in Tiled Pavilion, Istanbul Archaeological Museum]

Figure 2 *Osman Hamdi on the West Terrace at Mount Nemrud during Initial Investigations May 1883*, Topkapı Saray, Istanbul

yet resemble himself: "In the rare instances when a living person shows similitude with himself, he only seems to us more remote, closer to a dangerous neutral region, *astray* in *himself* and like his own ghost already [...] " (Blanchot 258). There is a resistance in Blanchot's argument to its own conclusions here, one that would be sustained by its apparent lack of concern with sexual difference and marked in the difficulties of his notion of the "neutral." The rarity of this effect of resemblance to self, this going astray, would be indicative of an inherited, historical determination of resembling, with women rather expected to resemble, while men assume the burden of apparently seeking to exclude the play of resemblance, at least unto death. The referents of these resemblings would not be stable, a critical point in this argument: women should resemble woman, as well as themselves; man should not just resemble man, let alone himself, but be, or at least continue to appear to be, the man he himself is supposed to be. But Blanchot's thought of resemblance, out of his argument about the image, cannot be contained by his lack of attention to questions of gender and sexual difference. The image is something that "sometimes serves the world's truth," but, at others, may be felt as that which "continues to affirm things in their disappearance," providing a "limit at the edge of the indefinite" that nevertheless does not always protect us from "the return of what does not come back, the strange heart of remoteness" that is the "life and the sole heart of the thing" (Blanchot 254-6). Or, presumably, people.

The corpse, then, provides Blanchot with a model of the strangeness of the image, as something that resembles something that no longer, for sure, exists. But, so that this argument does not fall back into an abyss of this temporality of loss, we need to follow Blanchot in his concern for the effect on a sense of place that the corpse and the image both engender. Where is the corpse? Because of the loss of the one who has died, the corpse engenders an effect of being out of place:

> Death suspends the relation to place, even though the deceased rests heavily in his spot as if upon the only basis that is left him. To be precise, this basis lacks, the place is missing, the corpse is not in its place. Where is it? It is not here, and yet it is not anywhere else. Nowhere? But then nowhere is here. The cadaverous presence establishes a relation between here and nowhere (Blanchot 256).

Funerary rites and the marking out of what Foucault termed the "heterotopia" of the cemetery, if not the colony or the museum, would be determined, according to Blanchot, out of a "fear of seeing *here* collapse into the unfathomable *nowhere*" (Blanchot 259). I shall return to Foucault's conceptualization of the museum and the colony later, in conclusion. But it is this relation between here and nowhere that can, I think, help us to understand the effects of images such as these photos taken of Osman Hamdi at Nemrud [figs. 2 and 3]: a sense of a placelessness of things and persons. The effect of the placelessness of even the most respectfully approached, judiciously considered and exhaustively studied and known archaeological remains *can* re-emerge. These photographs taken at Nemrud in 1883 can, via a consideration of the distinctiveness of their compositions, be understood to offer the possibility of an experience of a relation between here and the "unfathomable nowhere" (Blanchot 259), of which Blanchot writes, that accompanied and haunts the very reconstruction of the place of the archaeological finds.

A comparison of these photographs with other photographs of archaeological finds, sites and archaeologists taken in the nineteenth century may help to remark what is unusual about them, particularly in relation to what can usefully be identified as the norms of modern western European visual culture, on the one hand, and, on the other, the norms of a late Ottoman, if not more persistently Turkish visual culture on the other. Commenting on what he claims are the earliest known archaeological photographs, by Gabriel Tranchard, made on Victor Place and Félix Thomas' archaeological dig at Khorsabad in the 1860s, Frederick Bohrer has suggested a reading of photographs of archaeological sites in terms of "privilege" [fig. 4]. As Bohrer comments, deploying the studium/punctum distinction coined by Roland Barthes in *La chambre claire*, this first use of photography in connection with an archaeological dig is "not at all confinable to the normative *studium*":

> [...] this excavation space is treated almost as the backdrop of a stage, before which stand members of the excavation party, in various groupings. In one photograph, figures identifiable with the two French leaders of the excavation... appear in positions of privilege corresponding to those of the winged colossi [the Assyrian "*lamassu"*] (Bohrer 189).

Figure 3 *Osman Hamdi in front of the Head of a God, Mount Nemrud May 1883*, Topkapı Saray, Istanbul

What exceeds the "normative *studium*" or the search for period knowledge from photographs is, however, in effect a normative *punctum*. We should be struck, according to Bohrer, by the privileged positions adopted by the archaeologists, Victor Place and Félix Thomas, alongside the carved stone of the *lamassu*. This use of Barthes' distinction, however, fails to respect the traumatic character of the punctum, sublimating it into a methodology for the treatment of archaeological photographs. What Derrida termed the "haunting" (Derrida 41) of the space of the relation of our gaze to photographic images that Barthes's attentiveness to what wounds and captivates him in photographs brings out, in which what might just be a matter of a knowledge about time past is disturbed by a sense of something happening, here and now, and that seems not just to belong to there and then, is betrayed by Bohrer's objectivising methodological turn.

Figure 4 Gabriel Tranchard, *Gateway with Colossal Lamassu, Khorsabad c.710 BC* [with archaeologists Victor Place and Félix Thomas to far left and right], calotype, 1852-3

His account, then, does not explain his own deduction of the "privilege" of which this photograph is indicative: a gateway that had been excavated by the team under the French archaeologists, who, unlike their team, stand *outside* this newly exposed site. Reframing the Assyrian doorway, they offer themselves up as those who can frame a site, remarking the difference between its inside and its outside, even while, in their archaeological outfits, they appear to have risked appearing to disappear into the crowd of their more local team-members. The framing of the archaeological object by the authoring archaeologists, the compositional echoing of the theatricality of the doorway, is indicative of a boundary that would contain the threat to a sense of place that the discovery of another, past place, a place that is no longer a place, the archaeological "site," represents.

I may have begun to indicate to you why I think that the photograph of Osman Hamdi crouching down close in front of the head of Zeus-Orosmades betokens a relation to a sense of archaeological site that does not similarly repress its relation to placelessness, if not to an "unfathomable nowhere." Before I try to set this photograph and the other of Hamdi reclining and grinning on the head of Apollo-Mithras in relation to his earlier photography of the varieties of dress in the Ottoman empire, let me try to borrow some more of Bohrer's analysis in order to try to substantiate my points about the norms of modern Western picture-making and those of late Ottoman visual culture. Bohrer claims that photography has been used, by archaeologists, among others, "as a device for fragmentation or isolation of a single aspect from a culturally organic whole. Photography served […] as a way of leaving things out" (Bohrer 186). This romantic relation to representation and to place, both once complete and now fragmented, is born of a failure to explicate if not to understand his own reading of the "privilege" of the French archaeologists, a privilege that is also part of the heroic dedication of the Western archaeologist to their calling to unearth significant matter and to re-establish the boundaries that threaten a sense of future with an unknowable past as well as a sense of place with placelessness. As if in compensation for the claim that archaeological photography fragments the stable unity of a culture into knowable bits and pieces, Bohrer re-traces in tandem to the photography of archaeological sites an avant-garde concern with materiality and surface in work by photographers such as Félix Teynard [fig. 5].

Arguing that such an image offers up an experience of the materiality of surfaces, rather than the ideological work of the promotion of the heroism of archaeology, Bohrer misses the opportunity to expose what is in play here as a complex of identifications, projective and introjective, with the camera and with the surface of the image understood as materiality.

Figure 5 Félix Teynard, *Assouan, Arab Cemetery, Funerary Inscriptions*, Prints and Photography, National Library, Paris, 1852

He does not note the ideological work that this image probably served, in suggesting a certain disrespect for the space of the cemetery, either by Arabs or by their imperial rulers, the Ottomans. Instead, he deciphers the image as one that serves the emergent avant-garde sense of an image that, while it may represent something, is not first or foremost that. But this tactic, of seeking to draw attention to the surface of the image, by means of its composition, reminds us that this operates only as the disturbance of a norm that is operative from within Western visual culture. The previous photograph, by Tranchard, invited us into a projective identification with the space of the image, via the figures of the photographers in particular, as the site of a stabilization of the disturbance of place. The gaze insisted upon by photography is reterritorialized as a restoration of the site of archaeological activity as a localisable, reframable disturbance of a sense of place. In the case of the image by Teynard, however, at least for Bohrer, place would be forgotten in favour of an account of a disturbance of the veritable identity of the viewer, as materiality.

Introjective identification is theorized, in Freud, as an account of what has been proximate, but is, like the breast, removed.[1] But this norm of introjective identification forgets the anticipatory temporality that marks this example as a case of projective identification too, and which serves the imagining of the restoration of what is desired as what is lost. Thus, place might usefully to be rethought not simply as what has been most proximate, and which, now threatened with loss, must be defended, but also as a relation to the distant and the strange in the near. The history of Western encroachments on the Ottoman Empire in the name of archaeology, not to mention other rationales, suggests a repression of this within modern and modernizing Western norms.

It is perhaps this that Osman Hamdi understood and resisted, even while the photographs of him at Nemrud, among others, are indicative of their own norms of a sense of the distant and the strange in the near as a model of place. In order to support this contention concerning norms, and before I return to the photographs of him as archaeologist, I want briefly to consider his earlier ethnographic geo-history of Ottoman dress published by in 1873. *1873 Yılında Türkiye'de Halk Giysileri Elbise-I Osmaniyye* is a book of some 456 pages, divided into three parts, of people in dress from European Turkey, Ottoman islands and Asian Turkey, with respectively 23, 9, and 42 photographs. "Turkey" spreads beyond the bounds of its now current circumscription in so far as it includes people and dress from the "*vilâyet*" or provinces of Baghdad as well as Konya, Syria as well as Lebanon. Was Osman Hamdi more of imperial subject or more of a proto-Turkish nationalist? It would be significant to argue that there was, for him, no simple contradiction in being both imperial subject and nationalist, though this does not, in my view, entail a promotion of Ottoman culture as a model of co-existence, such as is implicit, if not explicit, in many current claims.

All the photographs were taken in Istanbul. Uniformly posed in front of a wooden panelled wall describing a part of a room, usually in threes, sometimes in twos, never on their own, the figures assemble as if passing through, in front of a gaze that would pass from the texts on the facing pages to the images and back again, via the calligraphic-like printings in Arabic above the frames of the photographic plates, and the names of the places from which the people come and to which, as it were, they will have gone back to, in Roman below, once they have all passed across in front of a gaze that would float somewhere apart from, if not above, some

---

[1] See "Introjection", Jacques Laplanche and Jean-Bertrand Pontalis, *The Language of Psychoanalysis*. London: Karnac Boks, 1988. 229-31.

ideally empty domestic space. Passing into view, for the time of the viewing of the reader of the book, the dressed body would be a model of what is expressive of the region that communicates via the imperial centre.

Figure 6 Plate 30, section 3, *Two Lebanese Christian Women and A Druze Woman* from Osman Hamdi, *Ottoman Public [or Popular] Costume in Turkey in 1873 [1873 Yılında Türkiye'de Halk Giysileri Elbise-I Osmaniyye*], Istanbul , 1873

I could pause too long over the remarkable photographic plate 30 of section 3 [fig. 6], of two Zahleli Christian women and a Druze woman from Lebanon. The commentary tells of the famous jealousy of the Druze men and of the belief, prevalent amongst the women, that sight is a kind of touch. As if not touched by the monocular look of the Druze woman, Osman Hamdi's text passes on to other people and peoples of the Hijaz, Yemen and Trablus *vilâyetler*, provinces or districts. Apparently untroubled in his scientific rationalism by such beliefs, it would be western commentators on photography such as Barthes who would claim the "contiguities" of touch suggested by the photograph, from light touching who or what is photographed, the photographic film and as if touching the viewer in turn. This ideal proximity of what is photographed, in the "wounding" punctum, across the reaches of time - the photograph, wrote Barthes, is the impossible conjunction of the "here now" with the "there then" - in inverting the western, positivist gaze that would make of every monument a document of its site of origin, making all photographs into quasi-documents of the failed remembering of site, resists, in its introjecting-projecting reterritorializing gaze, the distancing of light that is performed by photography.

Remembering Derrida's notion of the haunting of the gaze encouraged by photography, we can approach the relation in Ottoman visual culture between photographic representation of the people of the empire and an ideal of detachment from an ideal of domestic space. The people of the Ottoman Empire appear temporarily to occupy the spaces of this somewhat anonymous and grand domestic space. The space of the book contains the spaces of the photographic image, referring us to ethnographic particularities in either Ottoman Turkish or French (the book was also published in French), and in summary fashion, via the alphabets of the empire, to the identities and sites of origin, if not habitation, of these subject peoples. A projective identification with the camera position is encouraged in so far as the temporariness of the passing of these people through this domestic space is imagined. But there is also a more introjective identification with the same camera-position, in so far as it is with the technological means of the revealing of this passing through the once actual, but also imaginary ideal domestic-type space. This place, the site of the photographing of the representatives of the people of greater Turkey, is conserved as a site of both a projected and anticipated ideal of domestic space and a place in which the distant and the strange can appear: no simple contradiction between believing in the empire and in its different nations, the places, with their unknowable portion of the distant and the strange, that these appearances of people inhabit.

At Nemrud, then, such a reterritoralizing gaze was tested in relation to different objects, the massive remains of the hierothesion of Antiochus I. In one [fig. 2], Hamdi reclines, besuited, as if playing the part of the dandy-cosmopolitan, on the reclining head of Apollo-Mithras. The very comparison, between the reclining Hamdi and the reclining head, is suggestive at once of a sort of triumph of the archaeologist, whose head can still be raised, perhaps with a humour that would accept the eventual fall to come. I read this photograph as a sort of acting out of a relation to a way of relating to the power of representation to model: in the background, someone else is busy taking an impression or making a mould of a part of the carved stone at Nemrud. Adopting the contours of the massive head of the god, Hamdi invites an acknowledgement of a certain modelling after the figure, but not in its stony impassiveness. The smile would at once confess the power of representation and exceed it, in an expenditure of energy that, in turn, risks modelling the audience after its own humorous relation to the representative power of archaeological remains. The contrast with the photograph of Osman Hamdi crouching down in front of the head of Zeus-Oromasdes is striking [fig. 3]. Here the attitude appears to be one of puzzlement: but what would the object of that puzzlement be? Would it be to the identity of the found figure? Or to the identity of the finder, as it were, the puzzled archaeologist wondering where this head that is framed in this perhaps surprisingly close-framed image is to be placed or replaced?

Right hand to the side of his face, Hamdi sits. (Or does he squat? Is he even quoting the non-western squat?) Protecting himself from the fall of the sun, he may also have been obscuring his facial expression so as not to expose it to greater scrutiny. At the least, it appears as if he might have been posing as if to communicate a sense of puzzlement. We may not draw a line between the truly felt and the adopted, here, in relation to the question of the pose or to that of the power of representation. We might, further, wonder if the fez-wearing Osman Hamdi was encouraging or had been encouraged to imagine a comparison between himself and the apparently hat-wearing head of the decapitated god.

But it is the play in the relation to archaeological site, to place and placelessness, that is insistent, I think. It has been noted that the site of the hierothesion has not been reconstructed as it might have been. Rather than seeking to restore the heads to the bodies from which they had come, Hamdi's team set them in upright positions. To complain of this, however, would be to misunderstand the relation to place that was operative in late Ottoman visual culture. The hierothesion – a term coined so as to name this particular place – was not simply somewhere that ought to be restored to something approaching its original condition. There was no clear dividing line between places that were distant and strange, and places that were not. Imperial ambition of absolute territorial reach, this apparently contradictory tension may be momentarily exceeded if we accept the difficulty of both finding the remains of the different and the strange and not knowing how best to imagine their delimitation.

The foreground of the photograph of Hamdi squatting suggests the fall of the hill down which the head of the god has fallen, or been pushed. Above, the contours of the land as well as the outer edges of the figure of the head of the god disappear into the excesses of the light of the sun, as elsewhere they disappear into shadow. This effect of surface may be thought of as a sort of antitype of the modern Western avant-garde image of surface as materiality: surface as the site of the disappearance of appearances, as a limit of the informing powers of the image. Something similar might be remarked in other kinds of image-making in late Ottoman and in Turkish culture, including the paintings of Osman Hamdi. Instead of introjective identification with the materiality of the image, the incorporating identification would be instead with the means of this making appearances appear and appear to disappear: that is, with the camera and/or photographer or artist.

Such a relation to the powers of production would be part of an explanation of why Turkish culture has been so preoccupied with seeking to make room for technologies of all sorts, in excess of objective criteria of failures to produce or to advance. Such an explanation would have the virtue of indicating why difficulties and limits of technological power go apparently unacknowledged, when they do, given the play I am proposing here in this identification, between technical device and the model user of that device. I do not have the time now to defend this argument here, in terms that would suggest its being widespread, other than, perhaps, in so far as I have done so concerning the explanation of a sort of cultivation of an effect of surface in late Ottoman and Turkish pictorial culture.

My conclusion, however, concerning the ideal of mobility that is sustained by this relation to place as a site of the distant and the strange in the near, may serve further to render my analysis plausible. The remains found at Nemrud were rearranged closer to the open space of the hierothesion on the foothill of the Toros mountains in what is now south eastern Turkey. In contrast, the sarcophagi found in the garden in Sidon in 1887 were removed from there, and placed in the museum, under the directorship of Osman Hamdi, designed by the French architect Alexander Vallaury and opened in 1891. It has been claimed, by Wendy Shaw, that the housing of the Alexander Sarcophagus in the new wing in Istanbul demonstrates that: "While appropriating Western idioms of culture and heritage, the new museum building repositioned the people of the Ottoman territories as the most direct native descendants of the people who had made these objects and thus as the rightful heirs to the Hellenistic legacy" (Shaw 160). This would be true because of a newly-promoted sense of nativism, one that would wish to inspect its objects, of archaeological and artistic property, but also those very natives, as I indicated in connection with Osman Hamdi's 1873 work on popular dress, in a gaze that would float, introjecting the means of showing, be this camera, or the producing of the signs of Hellenistic idiom, or, as Shaw's phrasing suggests, the very building in which the finds were displayed. Shaw further argues that the early display of the sarcophagi stressed not an "evolutionary epistemology," but a geographic mapping of the sites of manufacture of various objects, what she calls a "stylistic arrangement," between the anthropoidal and the variously Hellenistic pieces. This was undertaken, I suggest, because of the distributions of the norms of identification encouraged by late Ottoman visual culture: crucially, the norm of a relation to places that would sustain them in their particular relations to the distant and the strange and which would link them in an impossible ideal of mobility that was, unlike the ideal of the mobility of modern, Western imperial cultures through a rational distribution of places in a homogeneous space, an instantaneity of transport from one place to another.

The cost of this would be the repression, between identifications with native and with the ideally mobile viewer, in late Ottoman culture of the very strangeness of the strange and the distance of the distant, including the "unfathomable nowhere" that Blanchot sensed as the undecidability of the truth of the image. There is a reproduction of the photograph of Osman Hamdi reclining on the head of Apollo-Mithras on display as part of the photo-text panels in a foyer-room between the anthropoidal and the Hellenistic sarcophagi excavated in Sidon and housed in the archaeological museum in Istanbul. The imperial museum represented the Ottoman colonies as parts that might hold together in a temporality of continuous instants, albeit ones haunted by the threat of placelessness that archaeological remains bring. The current museum has made room for a representation of a humour of modelling, if also of rule and rulers, though not for the corresponding image of a puzzlement at what to do about senses of place and placelessness. Foucault claims, in his fragmentary text "Of Other Spaces," in which he formulated the notion of heterotopias, that a museum, like a library, is an "immobile place" of "a sort of perpetual accumulation of time" (Foucault 234). The ways in which the ideals of a mobility through such an immobile place are interrupted by a sense of placelessness, however minimally, would tend to open them to being re-imagined. The introjecting incorporation of spaces and places of the dead, as the centrepiece of the Istanbul Archaeological Museum, would tend no longer to rule the sense of the place of the museum as somewhere to escape to or from.

Works Cited

Barthes, Roland. *Camera Lucida: Reflections on Photography*. Trans. Richard Howard. London: Jonathan Cape, 1982.

Blanchot, Maurice. *The Space of Literature*. Trans. Ann Smock. Lincoln, Nebraska: University of Nebraska Press, 1982.

Bohrer, Frederick. "Photography and Archaeology: the Image as Object." *Envisioning the Past: Archaeology and the Image*. Eds. Stephanie Moser and Sam Smiles. Oxford: Blackwell, 2005. 180-91.

Derrida, Jacques. "The Deaths of Roland Barthes." *The Work of Mourning*. Chicago: University of Chicago Press, 2001. 31-67.

Foucault, Michel. "Of Other Spaces." *The Visual Culture Reader*. 2nd edition. Ed. Nicholas Mirzoeff. London: Routledge, 2002. 229-36.

Shaw, Wendy M. K. *Possessors and Possessed: Museums, Archaeology and the Visualization of History in the Late Ottoman Empire*. Los Angeles: University of California Press, 2003.

Constantinople and the Slavs

R. M. Cleminson

Abstract

From the beginning the Slavonic peoples were in commercial and military contact with the centre of the greatest power in Eastern Europe, Constantinople, or Carigrad, the Imperial City. As they accepted Christianity, they became aware of and began to use other designations for the city: Constantinople, the New Rome. This paper investigates the relationship between Constantinople as a geographical reality and as a focus in the spiritual landscape which the Slavs inhabited, drawing in particular on pilgrimage literature, in which the two intersect.
This leads on to the question of "New" cities–Rome, Constantinople, Jerusalem–and whether it was possible for the Slavs to re-create Constantinople on their native soil. This is particularly relevant for those Slavonic realms which claimed to have achieved imperial status–*carstvo*–for themselves. It has particularly interesting manifestations in the competing ideologies of early Muscovy–not only in the famous theory of the Third Rome, but also in Patriarch Nikon's re-creation of sacred topographies.

> cum sint huc forsitan illa,
> haec translata illuc, summa tamen omnia constant.
> Ovid, *Metamorphoses,* xv, 257-8

From the earliest history of the Slavs, Constantinople was a major presence in their perceptions. The great city attracted trade not only from its Balkan hinterland, but also from distant Scandinavia, and the road from the Varangians to the Greeks lay across East Slavonic territory, with its two great entrepôts of Novgorod and Kiev at their northern and southern limits. The Slavonic peoples, particularly those who were close neighbours to the Empire, thus occupied a place in the great web of communications that surrounded the city in the early Middle Ages (Simeonova *passim*). Relations were not always friendly – the city was besieged by Krum in 813 and 814 and by Symeon in 913 and 924, by the Rus' in 860, and again in 941 and 944 under Igor'– but for the barbarians the city was not so much an adversary as an opportunity, and their expeditions as often as not ended with some commercial or political agreement. The city was, above all, the great power in the region, the pattern for all cities and the model for emulation.

The Slavs were thus familiar with Constantinople even before the historical record begins, and they called it Царьградъ, the Imperial City, as many of them still do.[1] This was evidently the earliest Slavonic name for the city, pre-dating Slavonic literacy, and duly appears in the very first Slavonic writings. The lectionary of the Codex Assemanianus contains eight commemorations of Constantinopolitan saints, for example on December 28th ⱄⰲ҃ⰰⰳⱁ ⱁⱌ҃ⰰ ⱀⰰⱎ҃ⰳⱁ ⱚⰵⰵⰴⰵⱃⰰ ⰰⱃⱈⰻⰵⱂ҃ⱂⰰ ⱌⱃ҃ⰻⰳⱃⰰⰴⱑ, "our holy father Theodore, archbishop at Constantinople" (f.135v). However, when as Christians the Slavs began to have a more intimate contact with the Empire and its literature, they also became more familiar with the Greeks' own name for the city, Κωνσταντινούπολις, and this is soon reflected in their own writings with the appearance of the form Константинъградъ, the City of Constantine.[2] An early example is found in Clozianus[3] 3a.24: ⱄⰲ҃ⰳⱁ ⰼⰵⰰⱀⰰ ⰸⰾⰰⱅⱁⱆⱄⱅⰰⰰⰳⱁ ⰰⱃⱈⰻⰵⱂⰻⱄⰽⱆⱂⰰ ⰽⱁⱀⱄⱅⰰⱀⱅⰻⱀⱑ ⰳⱃⰰⰴⰰ, "Saint John Chrysostom, Archbishop of Constantinople." This example is both early and typical, because it is precisely in this sort of context – the designation of an ecclesiastical dignity – that this name most frequently occurs. For some Slavonic writers, царьградъ and константиньградъ are evidently synonymous, and consequently the one may be substituted for the other in copying. (This occurs, for example, in the transmission of the *Vita Constantini.*) In other texts, however, there is a distinction – not absolute, but nevertheless a marked tendency to use константиньградъ in the titles of dignitaries and

---

[1] Even in those languages which call today's city *Istanbul,* the old name may still be used in a historical context.

[2] In both cases the word is translated rather than borrowed, and both parts, adjective and noun, are declined in the early texts (in fact one can find examples of Царьград declined in both parts in Russian as late as the nineteenth century); subsequently the adjectival part becomes fixed in the nominative-accusative form (e.g. Russian Царьград) or in the locative (e.g. Bulgarian Цариград).

[3] The *Codex Assemanianus* (Vatican Library, Cod. Vat. Slav. 3) and *Clozianus* (Trento, Museo Civico, MS2476) are two of the classical Old Church Slavonic manuscripts, and represent the earliest form of written Slavonic.

suchlike contexts, and царьградъ when the city figures as an actual geographical location. Though neither universal nor rigid, this distinction is both widespread and persistent over a long period of time and in various traditions. Thus in the lectionary of the Peresopnica Gospels (a Ukrainian manuscript written 1556-61) we find four archbishops of Constantinople commemorated, all described as костянтинаграда, while the translation of the image of the Saviour *to Constantinople* (August 16th) is expressed as въ царигра^д. Similarly, in a Serbian panegyricon written c.1580 and now in Budapest (Országos Széchényi Könyvtár, Quart. Eccl. Slav. 19), we find константиньградь consistently in the titles of the authors and subjects of the homilies (in most cases St. John Chrysostom), but in the heading of that devoted to the siege of the city in 626, вьнѥ́г^д а пєрси и҆ варварїи цр҇игра^д бранїѭ ѡ҆кроужишє.

Thus if константиньградъ establishes itself as the norm in official titles designating the spiritual dignity of the city, царьградъ continues to be used when the city is referred to in a historical or geographical context, as it very frequently is in mediæval Slavonic literature. It understandably looms large in historical writings, both original and translated, reflecting the Slavs' continued preoccupation with the Empire and its fate, and there are also texts specifically concerned with the city in its own right. It is especially these latter texts, where Constantinople is the subject rather than the setting for events, which reveal what the Slavs thought about the city itself.

Most of these accounts of the city were written by people who had visited it, for the benefit of those of their countrymen who were either unable to go there, or who were to come after. They belong to the tradition of pilgrimage literature, and some of them go on to describe Palestine. They are all East Slavonic in origin; the South Slavs seem to have lived too near to Constantinople to find it necessary to describe it in this way (though they do write about Jerusalem), but for the Russians Constantinople was as inaccessible as Jerusalem, and was similarly holy - and, unlike others, they preserved a very lively awareness of having received Christianity from Constantinople. The result is a vigorous tradition of pilgrimage literature which begins with Daniel of Novgorod in 1113-15 and continues to flourish into the nineteenth century.[4] In these texts, not surprisingly, the city is normally called царьградъ, though one of them, an anonymous text of c.1390, does provide what is apparently the first example of the city's Greek name in a Slavonic text, albeit in a

[4] It is in fact a valuable source for the study of the city; see Majeska 1984 and 2002.

curiously Slavonicised form and accompanied by a gloss: Аще кто поидет в Констянтинополе, в нарицаемыи Царьград.[5]

Constantinople is thus a sacred space for the Slavs, but there is very little sense of space in their descriptions of it. By and large these take the form of catalogues of churches and relics, which may occasionally be interrupted by the description of events. As a typical example we may take the description of his visit, made in 1348-49, written by Stephen of Novgorod. We can infer that Stephen was a member of the commercial élite of Novgorod, and that he visited Constantinople as part of a high-level delegation which was to discuss matters of church governance with the Imperial government. The point to be noted, however, is that we have to infer this: Stephen says nothing about it in his writings. We can also infer that he made a systematic tour of the holy sites of the city: by comparing the sites he mentions and the order in which he mentions them with what is known of the topography of fourteenth-century Constantinople, we find perfectly reasonable routes for each of the days of his pilgrimage. This, again, does not emerge of itself from the text: we have to bring external knowledge to our reading in order to make sense of his movements. For Stephen, and therefore for his readers (whose expectations he must be presumed to have taken into account, for he was neither an incompetent writer nor an unintelligent man), Constantinople is not so much the metropolis of the East as a sort of enormous reliquary, a cabinet of holy wonders. Only in the last sentence of his description does practicality suddenly burst through:

> А в Царьград аки в дубраву велику внити: без добра воже не возможно ходити, а иже начнешь ходити скудно или убого не можеши видѣти ни цѣловати ни единаго святаго, развѣ на праздники которого святого будеть, тоже видѣти ни цѣловати.[6]

It was difficult for him to include this vital information within the conventions of the genre.

Constantinople was thus a real place to the Slavs, but their image of it

---

[5] Majeska 129, 1984. Majeska believes moreover that this text was derived from a pilgrims' guide originally written in Greek. By and large the Greek form of the name does not appear in the Slavonic languages until the sixteenth and seventeenth centuries, under the influence of Western European languages.

[6] "Going to Tsargrad is like going into a great oakwood: you cannot go without a good guide, and if you go parsimoniously or without funds, you will not be able to see a single saint, except that on feast days you will be able to see and venerate the saint whose day it is." Majeska 1984, 45-47.

had little to do with actual topography, and much more to do with its symbolic importance as a centre of the Christian world. In paying great attention to this they do no more than follow the example of its inhabitants, for whom the city was no less than the New Jerusalem. This τόπος, which is common among Byzantine writers, may perhaps have originated as far back as Eusebius of Cæsarea:

> Καὶ δὴ κατ' αὐτὸ τὸ σωτήριον μαρύριον ἡ νέα κατεσχευάζετο Ἱερουσαλὴμ, ἀντιπρόσωπος τῇ πάλαι βοωμένῃ, ἣ μετὰ τὴν κυριοκτόνον μιαιφονίαν ἐρημίας ἐπ' ἔσχατα περιτραπεῖσα, δίκην ἔτισε δυσσεβῶν οἰκητόρων... τάχα που ταύτην οὖσαν τὴν διὰ προφητικῶν θεσπισμάτων κεκηρυγμένην καινὴν καὶ νέαν Ἱερουσαλὴμ... (*Vita Constantini*, III.xxxiii)[7]

Eusebius evidently refers to the re-foundation by Constantine of Jerusalem as a Christian city in place of the pagan Ælia, and the suggestion of the fulfilment of prophecy is altogether in keeping with the triumphalist tone of his work. It is however but a short step from Constantine as the builder of the New Jerusalem to the identification of Constantinople with that city. In later writers this is precisely what we find.[8] To take one example among many, John Mauropous, addressing Constantine Monomachus in his Second Oration on the Foundation of St George in the Mangana (1047), declares "πάντες οὗτοι συντρέχουσι ποσὶν αὐτομάτοις ἐπὶ τὴν Σιὼν τὴν ἁγίαν, τὴν πιστὴν μητρόπολιν ταύτην, τὴν Ἱερουσαλήμ σου τὴν νέαν"[9] (Lagarde 140).[10] The underlying idea here is not eschatological, but refers rather to Constantinople as the focus of the Christian οἰκουμένη – he has just called the Emperor "ὁ ποιμαίνων τὸν

---

[7] "And at the place where the Saviour suffered there was built a New Jerusalem, opposite the famous old one, which because of the murder of the Lord had been reduced to utter desolation, justice upon its godless inhabitants… perhaps indeed this was the second and New Jerusalem proclaimed in the prophetic oracles."

[8] See Podskalsky 204 for a short but representative list, from which, however, the oft-cited Epistle of Photius to Zacharias should be deleted as spurious ("Non potest Photii esse, nam scripta est lingua Armeniaca, sicut patet e stilo, e fontibus"–B. Outtier in Laourdas and Westerink, III, xi).

[9] "All these congregate at the holy Sion, this faithful metropolis, your New Jerusalem."

[10] An identical passage occurs on p.146.

ἐκλεκτὸν Ἰσραήλ"[11] – and is absolutely typical of Byzantine rhetoric, with its fondness for classical and Biblical archetypes. Several of the texts in which this comparison is made were among those translated into Slavonic in the early period (Stupperich 339), and the Slavs lost no time in imitating them, so that Kiev too becomes a second Jerusalem (Philipp 379). For the Slavs, however, the primary model for a Christian polity was of course the eastern Empire, and just as Constantine himself had joined David and Solomon as the archetype of the great ruler,[12] so Constantinople joins Jerusalem as an archetypal city. The adoption of Byzantine culture by the Slavs thus produces New Constantines and New Constantinoples, and even on occasion a New Thessalonica.[13]

The identification with Constantinople is not confined to rhetoric. It is no accident that the principle churches in Kiev, Novgorod, Ohrid and Sredec were all dedicated to the Holy Wisdom. The dedication of the Great Basilica in Pliska is unknown, but the prevailing opinion is that it was dedicated to the Mother of God, in which case its prototype is likely to have been the church at Blachernæ. St. Sophia in Kiev reflects its prototype in more than its dedication, and in fact a number of buildings in Kiev, both sacred and secular, indicate a conscious correspondence to Constantinople (Philipp 375-77). The Church of the Dormition in the Monastery of the Caves in Kiev deliberately refers to the church at Blachernæ, and indeed, according to the Patericon of the monastery, was built and decorated by Greek craftsmen from that very place. In its turn it served as the model for the church of the same dedication at Vladimir, which was itself the model for that in the Kremlin at Moscow.

Some caution is nevertheless required in evaluating the significance of these correspondences. In particular one should avoid rushing to the conclusion, common amongst writers too much in thrall to the dialectical

---

[11] "the shepherd of the elect Israel"

[12] The same phenomenon was also much in evidence among the Byzantines, and is examined in its different aspects in the various essays edited by Magdalino; the identification was immediately taken up by the Slavs (Vavřínek), and became so common that almost the same words can be used in eleventh-century Kiev of Vladimir ("Се вторый Костянтинъ в Руси явися", Nestor, *Чтение о Борисе и Глебе,* printed in Abramovič, 4) as of John Alexander in 1377 in Tărnovo ("дроугы ꙗвисѧ еже въ црѣхъ костан҄динъ," from the encomium on ff.311-312v of the Sofia Psalter, printed in Kodov 13). For the same identification applied to Serbian rulers, see Djurić 55-60.

[13] Поистинѣ Вышегородъ наречеся. вышии прѣвышии всѣхъ. вторыи Селоунь ꙗвися въ роусьстѣи земли – *Сказание о Борисе и Глебе* (Abramovič, 191).

concept of history, that comparison of one city to another necessarily implies a *translatio imperii.*[14] There is absolutely no basis for this, given that the Slavs had simply inherited the conventional use of archetypes from the traditions of Byzantine rhetoric. The temptation is particularly strong in the case of Constantinople, because, since Constantinople is the New Rome, any New Constantinople must, arithmetically, be the Third Rome, and commentators have been precipitate in identifying "New Constantinoples" as "Moscow the Third Rome"[15] in embryo, without considering that arithmetic and semiotics are very different sciences.

Thus when Metropolitan Zosima calls Ivan III "the new Emperor Constantine of the new Constantinople, Moscow" in the preface to the Paschalia compiled in 1492 (Tichonjuk 58-61), he is merely repeating the old formula, and not by any means promoting the idea of a third Rome that will take the place of the second, as has traditionally been assumed.[16] On the contrary, the whole point of this text is to emphasize continuity from the time of the Apostles to the present, and to dispel any idea that the Metropolitan is innovating. The first sentence sets the tone for the whole piece: "и якоже бысть в первая лета, тако и в последняя" (58) – as it was in the first years, so shall it be in the last.

It has become a commonplace of literary history that the idea originated with the description of Tărnovo as Нови Царьградъ in the Middle Bulgarian translation of the Chronicle of Constantine Manasses, and that this is was the model for the doctrine of Moscow the Third Rome which later became prominent in Russia. This is not borne out by a closer examination of the sources. The relevant passage is a comment by the chronicler immediately following his account of the deposition of Romulus Augustulus, which in the original reads:

Καὶ ταῦτα μὲν συμβέβηκε τῇ πρεσβυτέρᾳ Ῥώμῃ,
Ἡ δ'ἡμετέρα τέθηλεν, αὔξει, κρατεῖ, νεάξει,
Καὶ μέχρι τέλους αὔξειτο, ναὶ βασιλεῦ παντάναξ,
Τοιοῦτον σχοῦσα τηλαυγὴ φοσφόρον βασιλέα,

---

[14] For example, Maslennikova: "в псковской Хронографической Толковой Палее также выражена идея преемства царств. Автор пскович называет Киев вторым Иерусалимом" (203).

[15] The history of this notorious doctrine is far too complex to be treated fully within the scope of this paper, though it must of necessity be touched upon in any discussion of Constantinople and the Slavs. The best and most balanced statement of the classic doctrine remains that of Strémooukhoff.

[16] For example, by Lur'e 378-79.

Μέγιστον Αὐσονάνακτα, μυριονικηφόρον,
Κομνηνιάδην Μανουὴλ, πορφύρας χρύσουν ῥόδον,
Οὗπερ τὸ κράτος ἥλιοι μετρήσαειν μυρίοι.
Ἀλλὰ γὰρ πάλιν εἰς ὁδὸν ὁ λόγος ἀναγέσθω[...]. (*Compendium Chronicum* 2546-53)[17]

The Slavonic translation reads as follows:

И сїа оубо приклюѵишѫсѧ старомоу римоу· нашъ же новыи цариградъ доитъ и раститъ, крѣпитсѧ и омлаждаетсѧ, бѫди же емоу и до конца расти, еи царю въсѣми царствоужи, сицеваго прїемшоу свѣтла и свѣтоносна царѣ, великаго владыкѫ и изрѧднаго побѣдоносца, коренѣ сѫща іωана прѣизѧщнааго царѣ блъгарωм асѣнѣ, алеѯандра глаголѧ прѣкроткаго и милостивааго и мнихолюбиваго, нищїимъ кръмителѣ и великаго царѣ блъгарωм, егоже дръжавѫ слънца безѵислънаа да исѵьтѫтъ. **Н**ъ и пакы на пѫть слово наставлено бѫди (*Slavischer Manasses* 99-100).

This is usually understood to mean: "These things happened in the Old Rome, but our new Constantinople suckles[18] and grows, strengthens and rejuvenates itself, etc. [...]" and "our new Constantinople" is understood to be Tărnovo.[19] It is however immediately apparent that Tărnovo is nowhere mentioned by name in the Slavonic text, which moreover follows the Greek very closely, except for the omission of the reference to Manuel Comnenus in v.2551 and the substitution of the praise of John Alexander.[20] Most importantly, in the Greek text Constantinople

---

[17] "And these things happened in the Old Rome, but ours flourishes, grows, increases, renews itself, and shall grow to the very end, yea, universal emperor, now that it has received such a bright and radiant emperor, the great lord of the Mediterranean, many times victorious, Manuel Comnenus, the golden rose of the purple, whose reign will be measured by many suns. But let our discourse return again to its path […]".

[18] (Sic) The Slavonic доитъ results from a misunderstanding of τέθηλεν as if it were a part of θηλάζω instead of θηλέω.

[19] It is precisely in this sense that these "известни бележки" are understood in Tăpkova-Zaimova and Miltenova : "Това се случи в стария Рим, а нашият нов Цариград (т.е. Търново) цъфти и расте, укрепва се и младее" (48, 1988).

[20] This is not as incongruous as it might seem from a modern perspective. To the mediæval mind there was one Empire, just as there was one Church (and the two were largely coterminous), but since the days of Diocletian there could be more than one emperor, a tradition maintained at Byzantium (even if it rarely involved a genuine division of power). The Bulgarian rulers' title, "Emperor of Bulgarians and Greeks" (first assumed by Symeon in an era when there was still an Orthodox

is referred to not by name, but simply as ἡ δ'ἡμετέρα. The words нови and цариградъ are therefore glosses on the original, and in the absence of punctuation the words нашь же нови цариградъ could equally well be read as "our new one, Constantinople" or "ours, the new one, Constantinople." What little evidence we have of the mediæval understanding of the phrase points to precisely this reading: the same passage (up to v.2548) occurs in the Russian Chronograph, in which the phrase in question is further explicated as нашъ новыи Римъ, Царьградъ, "our New Rome, Constantinople," proving that the Russian compilers – who were working from the translation, not from the Greek – had no doubt about which city was meant.

Even more instructive is the use that the compilers of the Chronograph of 1512 made of this very passage as the model for their own comment on the fall of Constantinople in 1453:

> Сіа оубо вся благочестиваа царствіа, Греческое и Серпьское, Басаньское и Арбаназское и иніи мнози грѣхъ ради нашихъ Божіимъ попущеніемъ безбожніи Турци поплениша и вь запустѣніе положиша и покориша подъ свою власть, наша же Росиская земля Божіею милостію и молитвами пречистыя Богородица и всѣхъ святыхъ чюдотворцевъ растетъ и младѣеть и возвышается, ейже, Христе милостивый, дажь расти и младѣти и разширятися и до скончаніа вѣка. (*PSRL* 22, 439-40)[21]

It will be noted that no city is mentioned, nor any earthly ruler, and Russia is as yet only a land, not an empire, but we do see the germ of the idea of the metamorphosis of empire, of its transference from place to place, that will eventually culminate in the doctrine of Moscow the Third Rome.

---

emperor in the West), even though it implies absolute rule over a particular territory, is not inconsistent with continued spiritual pre-eminence of Constantinople as the metropolis of the entire οἰκουμένη.

[21] "Because of our sins and by the permission of God, these pious realms, the Greek, the Serbian, the Bosnian, the Albanian and many others were captured by the godless Turks, who laid waste to them and subjected them to their rule; but our Russian land, by the grace of God and at the prayers of the most pure Mother of God and of all the holy wonderworkers grows and rejuvenates itself and exalts itself, yea, merciful Christ, let it grow and rejuvenate itself and expand until the end of the age".

Similar ideas are developed further a little later in the century in the classic texts of the "Third Rome" theory, of the Epistles of Filofej[22] and the *Tale of the White Cowl of Novgorod,*[23] though neither of these mentions Moscow by name, and the *Tale of the White Cowl* is indeed generally considered to be anti-Muscovite in orientation. Here it is suggested that Russia was the only part of the οἰκουμένη that remained both free and Orthodox, and thus alone retained the imperial dignity. (It is notable that in some versions of the Epistles of Filofej the empire is not Russian but Romaic [Gol'dberg 75-78]). Both these texts certainly draw on South Slavonic sources, notably the Apocalypse of Methodius of Patara (Nikolov 93),[24] in respect of the kingdoms of the world coming together as (or being reduced to) one in the last days. It is only at the very end of the century, in 1594, that we find the first actual reference to Moscow specifically as the Third Rome (Ostrowski 175), and not until the seventeenth century that "Moscow the Third Rome" gained any real currency in Russia, as Križanić noted with some contempt: "Appellarunt Mosquam, tertiam Romam: et hoc Russia cum regnum ridicule fabulati sunt, esse Romanum: et huic regno dixerunt competere Insignia regni Romani" (Bessonov II 181-82).

These South Slavonic roots of the metamorphosis of empire lead us again into the cycle of Slavonic texts connected with Constantinople. In these texts it is the eschatological overtones that seem to find a particular resonance. Though full of real Bulgarian toponyms, they tend to avoid the name царьградъ in favour of константиньградъ, perhaps indicating a greater preoccupation with the mystical than with the historical. The *Bulgarian Apocryphal Chronicle* (Tăpkova-Zaimova and Miltenova 192-198, 1996) conflates Bulgarian, Byzantine and Biblical history and geography in the most extraordinary manner. The *Vision of Isaiah* (ibid 227-236) in the thirteenth-century Dragolj Miscellany (Зборник попа

[22] Edited in ПЛДР 1984 436-441. The textual evolution of these epistles is extremely complex, and it is possible that only one of them can be securely attributed to Filofej, the others being secondary texts derived from it. See Gol'dberg 92.

[23] Edited in ПЛДР 1985 198-233. Its dating is controversial.

[24] In respect of Filofej; this had already been noted by Strémooukhoff 96. Specifically, како снидошесе четири црьства во ѥдино (Methodius); вся царства православныя христианския вѣры снидошася въ твое едино царство (Filofej); вся христианская царства приидутъ в конецъ и снидутся во едино царство Руское (*White Cowl*).

Драгоља)[25] makes the traditional identification with Jerusalem – вь новѣмь ерⷭ҇лмѣ ꙗже зоветь се Коньстантинь градъ ("the New Jerusalem, which is called Constantinople") – but takes it even further: скрозѣ землю Изⷭ҇левоу. ꙗже се нарицеть Мѣзинаꙗ земли ("through the land of Israel, which is called the land of Mœsia")! Throughout the text the topography of Palestine and the Balkans is confused to such an extent that it is impossible to tell which of them is intended, or whether the place names have a purely mystical significance.

Like the old Rome, Constantinople is a concept as well as a place, *Roma mobilis,* capable of metamorphosis, capable of resurgence in different places. The Bulgarian literary fantasies referred to in the previous paragraph were hardly part of the official, imperial and Orthodox culture (and indeed the earliest cycle of them arose in the aftermath of the fall of the First Bulgarian Empire when there was no official Slavonic culture in Bulgaria) (Tăpkova-Zaimova and Miltenova 17, 1996). In Russia, however, once it began to aspire to the leadership of the Orthodox world, official culture began to look to a metamorphosis of Constantinople not merely as a model, but also as an authenticator of their own empire. The best symbol of this is the Cap of Monomach (шапка Мономаха), the diadem with which the Russian tsars were crowned from the first imperial coronation (of Ivan IV, in 1547) until that of Ivan V and Peter I in 1682. (The two half-brothers were crowned as joint tsars, the elder, Ivan, with the authentic regalia, and Peter with a replica. In 1721 Peter took the title *imperator,* thereby identifying himself with a Western European model of *imperium*, and the imperial regalia were re-designed accordingly.) According to a legend first recorded in the sixteenth century, the Cap of Monomach was given to Vladimir Monomach, grand prince of Kiev, by the Emperor Constantine Monomachus.[26] This is of course completely

---

[25] MS 651 in the National Library in Belgrade, containing a variety of texts, many of an apocalyptic or apocryphal nature and including the Apocalypse of Methodius of Patara referred to above; the *Vision* is Bulgarian in origin, but its presence in a Serbian manuscript indicates its wider currency over the Balkans.

[26] This legend received its definitive form in the *Сказание о князьях владимирских* (ed. Dmitrieva 171-178 and 185-191). It is noteworthy that the only historical personage associated with Constantine's gifts is Augustus, and that the text includes a genealogy tracing the ancestry of the grand princes of Moscow from the kin of Julius Cæsar which was considered sufficiently important to be incorporated into the coronation rituals of Ivan IV and Fedor Ivanovič (1547 and 1584). (This latter obviously lost its political significance with the extinction of the dynasty in 1598.) This preoccupation with Rome shows how far sixteenth-century Russia still was from the doctrine of the Third Rome as it would appear later.

untrue. When Constantine died in 1055, Vladimir was only two years old, and besides, the cap itself was made in the fourteenth century in the Islamic East, probably at the Golden Horde (Bobrovnickaja 63). After the Russians' first assertion of independence in the late fifteenth century, however, their Tartar political heritage rapidly declined in importance, and by the time their rulers claimed the imperial dignity their model was firmly Roman, that is to say Constantinopolitan.

This was to intensify in the next century, when Constantinople and its institutions became not only the archetype, not only the authenticator of the imperial power, but the model to which the Russian tsars increasingly sought to assimilate themselves, metamorphosing themselves into the image of Constantine and his city. In 1628 Michail Fedorovič had a crown made for himself in Constantinople "according to the pattern of the Emperor Constantine the Great" (Cicinova et al. 127). This was followed by a sceptre in 1638, and in 1660-68, under his successor Aleksej Michajlovič, by an orb and collar, all made with the same stipulation. The regalia of the Muscovite tsars were thus believed literally to reproduce those of Constantine.[27]

Byzance après Byzance perhaps, or perhaps not so much metamorphosis and place as a metamorphosis *of* place, a re-creation of the holy places within Russia. About the same time, with the full knowledge and support of the tsar, Patriarch Nikon founded three stauropegial monasteries in Russia. The first and greatest of these, not far from Moscow, known as the New Jerusalem, reproduces the topography of the Church of the Holy Sepulchre and its environs. The second, Iverskij, at Valdaj in the vicinity of Novgorod, imitates the Mone Iveron on Mount Athos, complete with an icon of the Mother of God Portaitissa specially made on the Holy Mountain. The third, on Kij Island in the White Sea (the same region as Solovki), is dedicated to the Holy Cross. Of its two churches, one is dedicated to the Exaltation of the Cross (September 14th), and the second to the Procession thereof (August 1st; the festival originated in the practice of taking the wood of the True Cross in procession from the Imperial Treasury to Hagia Sophia on that date). The chief object of veneration at the monastery was a full-size replica of the Cross of Christ specially made in Palestine and sent to Russia in obvious

---

[27] It is possibly these "insignia regni Romani" that Križanić had in mind. The collar and orb are illustrated in Cicinova et al. 126 and 129. The Byzantines had certainly encouraged the belief that the imperial regalia they used were those of Constantine the Great himself (*De Administrando Imperio,* §13), though since by the Emperor's own admission these are λόγοι πιθανοί it is doubtful whether this was actually believed at Constantinople.

imitation of the sending of the True Cross to Constantinople by St Helena. Russia thereby assimilates Constantinople as one of the three great sacred sites of Christendom.

The final stage of the metamorphosis is represented by the icon of the Adoration of the Cross painted in 1677-8 for the Church of the Crucifixion in the Great Palace of the Kremlin.[28] In the traditional iconography, Constantine and Helena have been joined by Aleksej Michailovič and his wife Mar'ja Il'inična and an unnamed bishop who is certainly Nikon; the texts they are holding are from the service for the Procession of the Cross, and the cross that they are venerating is unmistakably that of the Kijskij Monastery. In the semiotic framework of the icon, Moscow and Constantinople have at last achieved complete equivalence.

Works Cited

Абрамович, Д. И. *Жития святых мучеников Бориса и Глеба и службы им.* Петроград: Изд. ОРЯС имп. Акад. наук, 1916.

Бессонов, П. *Русское государство в половине XVII века. Рукопись времен царя Алексея Михайловича.* Москва: А. Семен, 1859-60.

Бобровницкая, И. А. "Шапка Мономаха: к проблеме происхождения формы." *Филимоновские чтения* 1 (2004): 59-85.

Цицинова, О. А. и др. "Тема греческого наследия в церковной и государственной идеологии XVII века." *Царь Алексей Михайлович и патриарх Никон.* Москва: Художник и книга, 2005. 103-129.

Djurić, Vojislav. "Le nouveau Constantin dans l'art serbe médiéval." *Λιθόστρωτον: Studien zu Byzantinischen Kunst und Geschichte. Festschrift für Marcell Restle.* Hrsg. B. Borkopp, T. Steppan. Stuttgart: Anton Hiersemann, 2000. 55-65.

Дмитриева, Р. П. *Сказание о князях владимирских.* Москва–Ленинград: Изд. АН СССР, 1955.

Гольдберг, А. Л. "Три 'послания Филофея': опыт текстологического анализа." *Труды Отдела древне-русской литературы* 29 (1974): 68-97.

Кодов, Хр. *Опис на славянските ръкописи в Библиотеката на Българската академия на науките.* София: Изд. на БАН, 1969.

Lagarde, P. de (ed.). *Iohannis Euchaitorum Metropolitae quae in codice Vaticano Graeco 676 supersunt.* Göttingen: Dietrich, 1882.

[28] Now in the Kremlin museums, inventory no. Ж-1714 and illustrated in Petuchova 133.

Laourdas, B. and L. G. Westerink (eds.). *Photii Patriarchae Constantinopolitani Epistulae et Amphilochia.* Leipzig: B.G. Teubner, 1983-88.

Лурье, Я. С. *Идеологическая борьба в русской публицистике конца XV–начала XVI века.* Москва–Ленинград: Изд. АН СССР, 1960.

Magdalino, Paul (ed.). *New Constantines: the Rhythm of Imperial Renewal in Byzantium, 4-13th Centuries.* Aldershot: Variorum, 1994.

Majeska, George P. *Russian Travelers to Constantinople in the Fourteenth and Fifteenth Centuries.* Washington: Dumbarton Oaks, 1984.

Majeska, George. "Russian Pilgrims in Constantinople." *Dumbarton Oaks Papers* 56 (2002): 93-108.

Масленникова, Н. Н. "Идеологическая борьба в псковской литературе в период образования Русского централизованного государства." *Труды Отдела древне-русской литературы* 8 (1951): 187-217.

Николов, Ангел. "Наблюдения върху цикъла старобългарски историко-апокалиптични творби от X-XI в." *Palaeobulgarica* 21.1 (1997): 91-107.

Ostrowski, Donald. "'Moscow the Third Rome' as Historical Ghost." *Byzantium: Faith and Power (1261-1557).* Ed. by Sarah T. Brooks. New York, etc., 2006. 170-79.

Памятники литературы Древней Руси: конец XV – первая половина XVI века. Москва: Художественная литература, 1984.

Памятники литературы Древней Руси: середина XVI века. Москва: Художественная литература, 1985.

Петухова, А. В. "Патриарх Никон и царь Федор Алексеевич. Икона 'Поклонение Кресту'." *Царь Алексей Михайлович и патриарх Никон.* Москва: Художник и книга, 2005. 132-35.

Philipp, W. "Die religiöse Begründung der altrussischen Hauptstadt." *Festschrift für Max Vasmer zum 70. Geburtstag.* Wiesbaden: Harassowitsch, 1956.

Подскальски, Герхард. *Христианство и богословская литература в Киевской Руси (988-1237 гг.).* Изд. второе, исправленное и дополненное для русского перевода. Перевод А. В. Назаренко под редакцией К. К. Акентьева. Санкт-Петербург: Византинороссика, 1996.

Симеонова, Лилиана. *Пътуване към Константинопол. Търговия и комуникации в Средиземноморския свят (край на IX—70-те години на XI в.).* София: Парадигма, 2006.

Die slavische Manasses-Chronik, nach der Ausgabe von Joan Bogdan mit einer Einleitung von Johann Schröpfer. München: Wilhelm Fink Verlag, 1966.

Strémooukhoff, Dimitri. "Moscow the Third Rome: Sources of the Doctrine." *Speculum* 28 (1953): 84-101.

Stupperich, R. "Kiev – das zweite Jerusalem". *Zeitschrift für Slavischen Philologie* 12 (1935): 332-354.

Тъпкова-Заимова, В. и А. Милтенова, "Идеята за християнския Рим и съответната терминология в късната средновековна българска книжнина." *България, Италия и Балканите: обществено-исторически и културно-естетически взаимоотношения XV-XX век.* София: Изд. на БАН, 1988. 45-50.

Тъпкова-Заимова, В. и А. Милтенова, *Историко-апокалиптичната книжнина във Византия и в средновековна България*, София: Унив. изд. "Св. Климент Охридски", 1996.

Тихонюк, И. А. "'Изложение Пасхалии' московского митрополита Зосимы." *Исследования по источниковедению СССР, XIII-XVIII вв.: сборник статей.* Москва: Ин-т истории СССР АН СССР, 1986. 45-61.

Vavřínek, V. "Slovanský panovník jako 'Nový Konstantin': Boris Bulharský – Rastislav Moravský – Vladimír Kyjevský." *Slavia* 70 (2001): 507-17.

IDENTİTY AND LOSS OF PLACE İN THE LOVE SONG: ON COSMOPOLİTANİSM, FORGİVENESS AND NEW CİTİZENS

CHARLOTTE TİNDALE SEXTON

Abstract

This article is part of ongoing research to honour the men who frequented a coffee shop in Fitzroy, Melbourne called *La Costa Azzura.* The themes of the paper are migration, communal estrangement, gender identity, ethics and kindness. The paper takes a trajectory through aspects of Turkish culture and history in the work of Diamanda Galás, Orhan Pamuk and Hussein Chalayan. Their work is critiqued in terms of Blanchot's poetics, Levinas's ethics and Derrida's late work on forgiveness and cosmopolitanism. This paper argues the work of these figures is relevant to forms of popular culture including rembetika music, Turkish psychedelic and *arabesk* music and Australian Rock musician Nick Cave.

I put a chain of love on my neck
I set an indistinguishable fire to my bosom Knowingly, I took the burden on my shoulders

Oh, beloved you took the word of a stranger
Sleep doesn't come to my eyes
I am a poor man abroad
No one looks at me
Come, beloved
I have sacrificed myself to you

You smashed, you crashed the crown of my throne You, again open my future I became unable to make my sorrow heard by you[1]

---

[1] Anonymous, Sevda Zinçiri (Melbourne, Melbourne International Festival, 2005), p. 14.

The Turkish love song “Sevda Zinçiri” was performed by Diamanda Galás in her work Defixions: Orders from the Dead at the Melbourne International Festival in 2005[2]. This work deeply offended Turkish nationalist culture because it was perceived as anti-Turkish. However Galas opens herself to Turkish history and culture by singing the Turkish love song. This song with uncertain origins tells the story of an encounter which still concerns the experience of contemporary global cities. Original separation and loss mark the experience of cosmopolitanism for minority existents without state or recognition. Darkness and obscurity is the condition of intimacy in separation and exile. For the author of the anonymous song existence is a burden and a chain without end. Writing is not the disappearance of distance but a hunger that nourishes itself in its remoteness and separation. The experience of writing is conditioned by radical exteriority that troubles the community of sameness. Maurice Blanchot sees the potential of poetic language to communicate to a community without certainty but a community beyond the anxious consciousness.[3] Their poetic testimony exceeds the logic which requires proof of existence, presence or recognition. Galás refers to the poets Baudelaire, Nerval and Pasolini as “my blood Brothers” (Galás in Zurbrugg 146).

Galás’s performances may be charged with being composed of women’s screams and foreign words that cannot be understood. Her work share’s with Antonin Artaud’s Theatre of Cruelty an ethical demand for a theatre of life rather than spectators (Artaud Antonin in Schumacher 108-09). The subject of this theatre becomes an active participant in a collective performance instead of being a passive viewer. Theatre is implicated in a spectacle of disease, madness, suffering and cruelty. Galas is the continuous historian less of her real history than of the symbolic events that have lead her body toward significance or threatened her consciousness with foundering. Galas enacted a personal catharsis and historical rupture concerning the trauma of the disaster that informed her childhood, the one that was passed down like a bloody relic from her grandfather to her father to her: “you don’t hear those stories for twenty

---

[2] Galás, Diamanda Defixions: Orders from the Dead (Mute, 2003); Galás, Diamada Defixions: Orders from the Dead and Songs of Exile (Melbourne, Melbourne International festival, 2005)

[3] Maurice Blanchot The Gaze of Orpheus and Other Literary Essays trans. Lydia Davis (New York, Station Hill, 1981) Maurice Blanchot The Space of Literature trans. Anne Smock ((Lincoln and London: University of Nebraska Press) (Melbourne, Melbourne International festival, 2005)

years and then forget them . . ." (Galás in Batchelder unpaginated 2005).The Historical incident of genocide is informed by familial memories even psychosis.

Certain expressions travesty orthodox culture because they are associated with the hated, despised and illegal undercurrents of society. The author of "Sevda Zinçiri" expresses the irreducible anonymity and essential solitude of the community of individuals destitute in their mortality-a community of those who have nothing in common. The song expresses what was originally repressed and absent from history, from modernity and from the city. The archaic song expresses an ethic concerned with cosmopolitanism, forgiveness and new citizens in contemporary culture which preoccupied the work of Jacques Derrida. He argues that cosmopolitanism and forgiveness exceed the politics of tolerance. The condition of global cosmopolitanism afflicts many cultural subjects with original separation and loss. The ethical implications of cosmopolitanism are present in Orhan Pamuk's Istanbul memoirs and Hussein Chalayan's fashion projects.

Galás's work was indeterminate because it traversed poetry, performance, art and history. The work was composed of texts expressing the polyphonic voices of a history "written from the margin for the margin, the living margin . . . .the people who have been isolated or kept aside by the larger part of society . . . the Invisible (Galás in Zurbrug 155)".1 Galás's work expresses the abjection of those who are not the same and judged as if they should desire to be like and the same. She resists the abjection of the victim in spite of their unrelenting suffering and oppression. In response to this trauma and its ethical demands Galas produced works for marginalized communities including homosexuals, prisoners and psychiatric patients. Defixions are curse tablets placed on graves to protect the dead. They were expressions of low culture associated with Asiatic Greeks. Yannis Ritsos writes: "We should guard our dead and power, lest by some hour our Opponents disinter them and take them away."[4]

---

[4] Yannis Ritsos The Graves of our Dead trans. Kimon Friar and Karelisa Hartigan, N.C. Germanacos, Thanasis Maskaleris, Kostas Myrsiades and Gwendolyn MacEwena and Nikos Tsingos in Galás, Diamanda Dexions: Orders from the Dead 5Various Love Peace and Poetry: Turkish Psychedelic Music (Germany, Q.D.K Media, 2005) http://psychevanhetfolk.homestead.com/TurkishProgressive.html <http://psychevanhetfolk.homestead.com/TurkishProgressive.html>
http://www.onstagerecords.com/Turkish_psych/turkish_psychedelic.htm
<http://www.onstagerecords.com/Turkish_psych/turkish_psychedelic.htm>

The work Defixions: Orders from the Dead included texts written by Ritsos, Atom Yarjanian, Ali Ahmad Said, Freidoun Bet-Gulawi and Paul Celan. Poetic writing and speech is addressed to the other, a community of existents altered through existence3. Writing is a powerful ally of the disinherited provided that it is accepted by another. However Galás addresses a kind of genocide which is becoming increasingly enacted in a generalized and all pervasive condition: You die because you are the face of the future. The trauma and repression of genocide involves a destructive hostility directed at what Levinas calls the face of the other, the similar other, the closest neighbor (Derrida 2001,50) The anonymous Turkish speaking author expresses the condition of the other afflicted by exile from the classical and metaphysical condition. This is a condition of proximity rather than distance which also afflicts the Jew in Levinas's work .Galás's work is ethically placed in relation to Levinas's ethics of the other and the catastrophe of metaphysics which then influences Derrida's appeal to forgive the unforgivable. Forgiveness concerns "the universal urgency of memory" (Derrida 2005, 28). Forgiveness "should remain exceptional and extraordinary, in the face of the impossible" (Derrida 2001, 32).

Seeking an idealized other is to risk the break up of speech in psychosis and subjective death. This marked the decline of the poet Nerval who hung himself in an Istanbul street. The poet wrote despairingly in "The Disinherited": "I am saturnine, bereft, disconsolate . . . The flower that so pleased my distressed heart (Nerval in Kristeva 141). The phantasm of the anonymous love song "Sevda Zinçiri" is a feminine figure. The author writes to give her form and reality. The lonely migrant male commemorates the lost thing or object. He is witness to the ethical kindness of the stranger who offers the gift of respite and transient community. Love, compassion and forgiveness, aging, memory and kindness become the essences of life and art as gratitude directed toward the other. These are the essential experiences of life and art. We love the other who understands the source of our suffering, the place of our passions, our desires, our prayers and our daily vocations. The woman in the love song is a strange and anonymous woman; a foreign woman but not necessarily from elsewhere. Such a woman may be seriously affected, ill or intoxicated but she loses nothing of her material degrees of freedom. This is a disenfranchised female. She is marked by her separation from the mainstream of feminism. She is singular, alone and stigmatized by normative and legitimate social institutions. She is a travesty of official

---

I was introduced to the folk pop singer Burcin by my Turkish language teacher during Turkish Language classes at the Council of Adult Education in Melbourne.

authorized culture. She risks being victimized and propelled into "a narrative of inexorable decline and fatality, from which she cannot disimplicate herself" (Sedgwick in Kwinter 582). She is at risk of being unable to author her own identity by the institutional discourses of the law, feminism and medicine. This feminine subject is at risk of entrapment and victimization. She risks being disempowered of the right to respect and dignity of the universal subject

She is installed as the proper object of compulsory institutional disciplines, legal and medical, which without actually being able to "help" her, nonetheless presume to know her better than herself-and indeed, offer everyone in her culture who is not herself the opportunity of enjoying the same flattering presumption (Sedgwick in Crary and Kwinter 582).

For minority communities love is experienced in relation to estrangement and strangeness rather than the safety and familiarity of the same. This is a condition of Orphic space. The tale of Orpheus concerns the attempt to reach the dead and "to gain from death the possibility of speaking and the truth of speech" (Blanchot, 41, 1981). Blanchot became preoccupied with poetry and Orphic space as a condition of poetic writing marked by original separation and loss. The myth of Orpheus and Eurydice is of originary and archaic mourning that is endemic to all speaking beings. Orpheus descended to the underworld to retrieve Eurydice from premature death on condition that he never gazes upon her again. He failed to respect this limit, lost her and was then torn apart and destroyed. Eurydice dwelled in nocturnal darkness a figure whose strangeness excluded all intimacy. Orpheus makes recourse to art through the loss of intimacy. The gaze of fascination warns of separation and loss rather than longed and hoped for unity and intimacy. What happens when their eyes meet? Orpheus passed beyond the law that forbids him to return. Eurydice does not exist except in his glance back. The glance of the gaze becomes a beautiful image related to the experience of mourning. Orpheus's gaze is touched by the night and inscribed in a body whose secrets belong to the night. The lovers are ghosts in nocturnal space. The gaze prefigures her disappearance. In Orphic space desire is absolute if the desiring being is mortal and the desired invisible. Writing becomes the impossibility of finding one's place. The disappearance of the figure of the other is linked to a name that cannot be owned or recognized. The figure becomes a stranger to oneself. Orhan Pamuk writes about the orphic relationship between love, loss and memory in *My Name is Red*:

> Love, however, was a distant and forgotten thing, like my memories of having lived in the city. . . . I was slowly forgetting the face of the childhood love I'd left behind. With growing panic, I tried desperately to

> remember her, only to realize that despite love, a face long not seen finally fades . . . I knew that the face I imagined was no longer that of my beloved . . . I was fully aware that my beloved's face had long since escaped me (Pamuk 2001, 6).

The feminine figure in Orphic space is obscure, veiled and ambiguous-a woman who dwells in hell. She simultaneously appears and withdraws, disappears and vanishes. She is fascinating because she menaces and seduces. Her image is utterly attractive and terrifying in the face to face encounter. This is the figure of style which affects Galás's image and performances. Similarly writer Kathy Acker is drawn to Eurydice as a figure who is intimate with the underworld: a space which figures in art and poetry. Acker writes that Eurydice was a woman who began to search for who she was and alone and solitary kept abandoning and returning to love. Identity begins in the shadowy night of being's relation to death. Acker writes:

> Every Angel is Terrifying
> Because of this, because I have met death, I must keep my death in me, gently and yet go on living.
> Because of this, because I have met my death, I give myself birth. (Acker 187-88)

Eurydice is a woman who cannot speak, though she finds ways to speak otherwise-silently through her solitary appearance. Some figures make such an impression without speaking and in spite of the brevity of our encounter with them. These figures prolong finite existence. They are preserved in our memories and dreams long since they have departed. The necessity of art reveals minute perceptions, hallucinations, distractions and reveries. Art makes the absent present and the imperceptible visible. Art memorializes the absent community.

Australian rock musician, Nick Cave argues that the sadness of love songs is related to the limit of violence and non violence present in Frederico Garcia Lorca' lecture 'The Theory and Function of Denude' and the of Portuguese experience of inexplicable longing saudade (Cave 7). The love song expresses the haunted premise of longing and separation experienced by the child, the lover, the lunatic and saint. The song becomes the unique gift of language. Cave argues that a love song acquires its own identity, its own life and its own truth. The love song is a unique instant addressed to the future. As long as there is time there is hope and the possibility of beginning a new. The sense of the song's past derives from its sense of the future. The promise of the future is a promise of resurrecting the past, with all its forces, but in such a way that it can begin anew. He writes:

> . . . the peculiar magic of the Love Song, if it has a heart to it, is that it endures where the object of the song does not. It attaches itself to you and together you move through time. But it does more than that for just as it is our task to move forward, to cast off our past, to change and to grow, in short to forgive ourselves and each other, the Love song holds within it an eerie intelligence all of its own-to reinvent the past and to lay it at the feet of the present" (Cave 3).

The language of the other contains a trembling that can be heard, without the discreet tremor of language that undersigns. The love song concerns a language of origin and a subject which has lost his language and his homeland. He speaks in another language because he is at risk of losing his mother tongue. Speech and art do not let the slightest disturbance appear on the surface. In this sense migration and exile is underscored by "monolingualism of the Other" And "the prosthesis of origin", a question of linguistic and cultural identity which preoccupied Derrida when he wrote: We only ever speak one language. We never speak only one language (Derrida, 8, 1998). The solitary soul dwells in a language that one will never possess but which one is destined to speak. Orhan Pamuk's memories of Istanbul integrate its cosmopolitan history and what is happening to its history. He writes: "In 1852 . . . you could hear Turkish, Greek, Armenian, Italian, French and English (and more than either of these two languages, Lad-O the medieval Spanish of the Jews who'd come to Istanbul after the inquisition-many people were fluent in several languages" (Pamuk 2005, 215).

When citizens move from one place to another their history does not count as history to many because it is composed of detritus and trash such as melancholy, sentimentality and nostalgia. In rembetika music the immigrant is condemned to speak a language that will never be his. Suffering and pleasure exist in the condition of speaking a language which is not one's. Rembetika music is of uncertain origins but has pronounced connections to Middle Eastern music styles which migrated to Greece with the influx of refugees (prosfigakia) from Asia Minor. Many of the refugees had lived for generations in Turkey, spoke Turkish as their first language, and had acquired Turkish ways of life. The term 'rembetika' includes Turkish terminology meaning 'outlaw' or 'outsider' and Greek terminology meaning 'unsteady' and 'unpredictable'(Holst, 13: Kritsos, 1-3). This style of popular-folk music emerged from the underworld of cafes amans. The café aman was a space where cultural expression came from below. It was a space of isolation, alienation, violence, crime, sex, excess and delirium. These spaces and those who frequented them were stigmatized and denigrated as criminals, the poor, the unemployed and drug addicts. This was a minority society of male outsiders and outcasts. An outsider community is familiar with the isolation and marginalization that characterizes modern alienation in its extremes. The Rembet was a social type at risk of being expelled from history. The affirmation of maligned difference and otherness is an ethic of hatred related to the politics of difference and the ethical struggle of the other. The ethic of the outsider can be transmitted and intuited between artists and historical epochs in expressions of suffering and outbursts of pleasure.

Amans are a vocal improvisation (taximi), usually in the beginning of the rembetika song. Amans used in rembetika music are both occidental and oriental and can also be heard in Turkish, Romane (Gypsy), and Kurdish music. The messages of the rembetika songs were often the harsh, selfish and poorly written works of an illiterate subculture. The expression of difference and identity in rembetika music was related to the non-mastery of an appropriated language. Amans express deep emotions and suffering of the outlaw, the outsider who is suffering, wronged and brutal. The rembet also transformed pain, suffering and idle tragedy into a lifestyle of pleasure and beauty antithetical to the bourgeois subject. In the dream of the café aman: The underworld manges would sit on the floor around a charcoal brazier while a boy filled the narghilé with Turkish hashish and passed it around. During the song, one of the manges, already high, might get up slowly to dance. As he circled and swayed, the others smoked and sang, while their harsh life was forgotten (Holst, 19-20). For the rembet, affections are experienced through the body in alcohol, gambling, drug use, prostitution or mysticism. Intoxication and delirium were present in their trance-like dances and music. The manges were identified by their style of dress associated with anti-social behaviour. This was a kind of masculinity which alternates between femininity (beauty, mysticism and lyricism) and stoicism. The café aman was a world of extremes and excesses revolving around binary cultural terms which Derrida attempts to critique in the ethics of Cosmopolitanism: Europe/Orient, exile/migrant, city/countryside, sacred/profane, male/female. The liminal and polyphonic culture of the café aman was mystical and touristical, extremely tough and rigorous. The rembet came from the other and remained with the other as they could never return home. The café aman became a site of oriental fold; a full and void space of ephemeral movements and encounters between others and strangers. Its community was improper, excessive, eccentric, extravagant and duplicitous.

This predominantly masculine cultural space is stigmatized by perceptions of misogyny and potential violence against women. Masculine domination is expressed in performative rituals as compensation for social marginalization. Masculine domination is also expressed in the "gross flattery" and "shrewdness" (Pasolini, 57). Male domination is "a gentle violence, imperceptible and invisible even to its victims, exerted for the most part through the purely symbolic channels of communication and cognition . . ." (Bourdieu, 2) Pierre Bourdieu argues that "ritual stereotyping", "phallonarcissistic vision" and "androcentric cosmology", "are common to all Mediterranean societies and which survive even today, but in a partial and, as it were, exploded state, in our cognitive and social structures". This argument implies that women were victimized and disempowered. The women who frequented the café amans were referred to duplicitly by the Turkish word hanoumi meaning both lady and whore (Holst, 48-49). The experience of women in such spaces includes the potentiality of the profound empowerment and vulnerability that characterizes the liminality of heterosocial spaces. Women who occupy these spaces are stereotyped victims of male violence, whores or self-abusers. In Albania, women have the mobility to visit the coffee shops and tea house in the guise of "sworn virgins" or tobija (bound by law). The sworn virgin is a woman who lives as a man. She may be reared by her family as male or reconstruct herself as male in later life. The muskobanja is a man-like woman. These women live as men and perform male roles. In this custom they do not take partners or families nor have children. (Grémaux, 244)

The love song which emerged from the culture of café amans and rembetika music expresses the melancholic quality of hüzün which the writer Orhan Pamuk argues is the ethic of life lived in Istanbul24. History becomes the hope of altering time otherwise, of returning and bringing back. Orhan Pamuk discusses the Turkish word Hüzün which accompanies illness, poverty and afflicts the addict. The resignation to melancholic existence involves a politics and ethics by embracing the anonymous struggle of failure, intoxication, defeat and poverty. Hüzün teaches endurance in times of unrecognized poverty and deprivation. Hüzün concerns a "life that endures death and maintains itself in it" (Blanchot, 41, 1981). This is the obscure existence of life deprived of light and recognition. Ordinary speech begins with the rhyme of anonymous survival and perseverance: an incessant murmur of that which is endlessly suffered and endured-turned into another kind of time. Hüzün is also approximate with intoxication and stoic beauty. Hüzün is not "an illness for which there is a cure" or an "an unbidden pain from which we need to be delivered" but is a way of life which is freely chosen (Pamuk, 93, 2005). It is a communal rather than solitary anguish for what has been lost. It is an impoverished style associated with worldly pleasures and material gain. Alternatively, in Sufism, it is the anguish we feel through our separation from Allah (Pamuk, 81, 2005) Alienation, failure, illness and grief are affects expressed in sentimental and popular forms including Turkish pop songs such as Burcin, Gesi Bağlari and Turkish psychedelic rock such as Sonra's Selda Bundun and Ince Ince Bir Kar Yağar5. Popular music has the potential to produce physical pain and grief in the listener associated with absence, separation and loss. Melancholic existence is turned toward the "elsewhere", the "otherwise" and the "other".

The café aman was a kind of passage like the arcades described by Walter Benjamin. He argued that melancholy is an affective response which also accompanies religious and political crisis Kristeva 171). Imagination becomes a response to deprivation. The other is stigmatized to suffer the mournful melancholy of unhappy consciousness. Melancholy is expressed in phantasms of memory and perception. Internalized figures from the past become present in dark forms more beautiful than a dream. The displacement of migration, exile and failure are marked by the pain of transience and memories of the other. The ethos of such a space is expressed in a wish to escape from time forever. The café aman can enable this as it is a place where space becomes time and time stands still. The café aman is an acute site of mortality and fate's misfortunes, a kind of memento mori described by Walter Benjamin:

> Everything about history that, from the very beginning, has been untimely, sorrowful, unsuccessful, is expressed in a face-or rather in a death's head . . this is the form in which man's subjection to nature is most obvious and it significantly gives rise not only to the enigmatic question of the nature of human existence as such, but also of the biographical historicity of the individual (Benjamin 166).

The affirmation of poverty and the struggle of community in deathly melancholy are present in Charles Baudelaire's poem The Liturgies of Satan. Baudelaire intuits the condition of exile and affirms the condition of negativity which is repressed, expelled and silenced. Baudelaire wrote poems in the café or on the street. The poet participated in revolutionary struggle and he experienced a decline from aristocratic elegance to poverty and illness.

> Father to those whom in his sombre wrath God drove from his Paradise on Earth,

Satan have mercy on my long distress! (Baudelaire in Richardson 194)

The ethic of Baudelaire's The Litanies of Satan (also performed by Galas) migrates, displaces and travesties perversely toward a good that is the other of cultural appropriateness. A good requiring risk and courage as it loses itself in the extremes of pleasure and pain: Satan have mercy on my long distress!18 Baudelaire is ethically complicit with the world of darkness and despair inhabited by minorities. Artifice and spleen involve processes of inversion, underlying violence and ecstasy. The lost soul is internalized in a figure more beautiful than a dream. Poetic language is subjected to the undecidability of real and imaginary. Figures of the past are phantasms of perception, imagination and memory which risk one's integrity and truth.

Galas's performance of the love song, "Sevda Zinçiri" and its uncertain origins in Turkish rembetika cultures are implicated in the ethics of cosmopolitanism. The history of the present includes a terror which afflicts so many new citizens who become layered with personal stories and burdened with what passes for collective memory. Ethics is an act of forgiving the unforgivable. Terror and trauma are residues of history which does not count as history because it is composed of detritus and trash. What becomes of those who are dispossessed of being and victimized into silent self abuse? Derrida's translator Spivak interrogates elite cosmopolitanism and alterity in institutions which produce history. "And I must keep repeating, remaking history is a tall order . . . 8Spivak's statement has implications for reconciliation and forgiveness in a personal and historical sense. Genocide is an ethical concern of contemporary cosmopolitan societies. Cosmopolitanism exists in relation to the homogeneity and hegemony of cultural language and forms. Derrida argues cosmopolitanism is ethically involved in the work of forgiving the unforgivable: "forgiveness without power" (Derrida 59).

The new cosmopolitan citizen concerns the figures of a past that was originally repressed. Pamuk writes: we live in an age defined by mass immigration and immigrants" (Pamuk 2005, 5). Fashion designer, Hussein Chalayan interrogates his privileged cosmopolitan citizenship in relation to his Turkish origins and the trauma of the conflict between Greece and Turkey in Cyprus. Textile patterns become a language of migration in Chalayan's work. The designer anglicized his name when he moved to London to study: Calayan became Chalayan.[5] Turkey's attempts to modernize include the repression of its Ottoman history. The Young Turks

[5] This was suggested to me by the proprietor of Burger Culture in Williamstown, Melbourne.

went so far as to ban Ottoman clothing and legislate for the compulsory wearing of modern clothes. Chalayan explores the relationship between modern dialectical progress and forms of dress. His collection Geotropics (Spring/Summer 1999) (Evans et al. 52-61) involved an exploration of the relationship between cultural tradition, nationalism and the garment through processes of technological morphing. In his collection Anthropology of Solitude (Autumn/Winter 2004) (Evans et al 158-165) fabrics included images drawn from Turkish modernization in the 1920s. The basis of a nationalist consciousness required the construction of a common history through propaganda images of modernization. His collection Ambimorphous (Autumn/Winter 2002) (Evans et al. 120-127) involved a dualistic relationship between traditional Turkish clothes and the form of the black dress. The evolution of the garment is not organic but is fragmented. The designer explored the reductive modernist programs to eliminate ornament and decoration. Chalayan's garments indicate traditional cultures do not exist outside the contemporary economy and are experiencing a new mobility and sophistication enabled by migration, technology and the media. The designer implicates the effects of these processes on intergenerational memory and connection to place.

Turkey's transition to modernity involved a mythical transition from a nomadic culture to a sedentary culture. The hans or caravanserai of Anatolia were built by the Seljuk Turks at regular intervals along their trade routes. In contemporary Turkey they exist as memento mori and tourist relics. Modern Turkish culture then judged roads to be corrupt spatial mechanisms34. In Chalayan's video installation Place de Passage (2003) (Evans et al, 180-183), the modernist figure of the flâneur becomes a feminized figure that travels in a pod from London to Istanbul. She expresses a new subjectivity which is mobile and in constant transition. The pod enables self-sufficient autonomy and virtual mobility. The figure is in perpetual motion as the pod glides over the Bospherous. She never stops or rests. This figure is also without home or origins: "that from which and by which something is what it is and as it is" (Heidegger in Cazeaux 80). Afterwords (Autumn/Winter 2000) (Evans et al. 80-89) acknowledged that movement and mobility is experienced without choice and without hope. Refugees flee their homes and countries at times of war and political persecution including forms of censorship, torture and genocide. Accompanied by a Bulgarian female choir, models were dressed in simple under slips. They then removed the covers from chairs in a living room and wore them as garments before leaving the imaginary living room in which they were situated. The furniture in the room was folded up to function as suitcases and garments necessary for movement and migration. The collection concluded with the sense of absence experienced by the migrant and refugee in affects such as memories, isolation, silence and melancholy. Observations were made between this collection and the Kosovo crisis that marked Balkans war.

Love songs are residues cast off in the search which accompanies exile. The songs are like companions offering protection and comfort. While the songs express loss, longing, madness, melancholy, ecstasy, magic and joy they are also testimonies of respect and gratitude. These sentiments exist and survive in silence. The love song "Sevda Zinçiri" expresses an ethic which values sadness. This emotion is not to be medicated as its suffering teaches empathy and respect. Trust and risk require one to face another who both appeals and contests - "existence is destined for this other". The extreme experiences are the events for which we are most grateful. Silently, the watcher longs for the lost soul that they will never see again but whom they remember with gratitude. This sad but plentiful silence expresses love not recognized because it is the mark of an absence that is not lack.

Works Cited

Acker, Kathy. *Eurydice in the Underworld.* London: Arcadia, 1997.

Blanchot, Maurice. *The Gaze of Orpheus and Other Literary Essays.* Trans. Lydia Davis. New York: Station Hill, 1981.

—. *The Space of Literature* trans. Anne Smock. Lincoln and London: University of Nebraska Press, 1982.

Benjamin, Walter. *The Origin of German Tragic Drama.* London: Verso, 1998.

Bourdieu, Pierre. *Masculine Domination*. London: Polity, 2001.

Cave, Nick. *Complete Lyrics*. London: Penguin, 2001.

Cazeaux, Clive, ed. *The Continental Aesthetics Reader*. London and New York: Routledge, 2000.

Crary, Jonathan and Sanford Kwinter, eds. *Incorporations.* New York: Zone, 1992.

Derrida, Jacques. *On Cosmopolitanism and Forgiveness.* London and New York: Routledge, 2001.

—. *On Monolingualism of the Other or The Prosthesis of the Origin.* Stanford: Stanford University Press, 1998.

Evans et al. *Hussein Chalayan.* Rotterdam: Nai and Groninger Museum, 2005.

Galás, Diamanda. *Defixions: Orders from the Dead.* Melbourne International Festival, 2005.
—. *Defixions: Will and Testament.* Mute Records, 2003.
Grémaux, René. "Woman Becomes Man in the Balkans." *Third Sex, Third Gender: Beyond Sexual Dimorphism in Culture and History*. Ed. Gilbert Herdt. New York: Zone Books, 1996.
Harrison, Charles and Paul Wood, eds. *Art and Theory 1900-1990: An Anthology of Changing Ideas*. Oxford and Cambridge: Blackwell, 1992.
Holst, Gail. *Road to Rembetika: Music from a Greek Subculture: Songs of Love, Sorrow and Hashish.* Athens: Anglo-Hellenic Publishing, 1975.
Kristeva, Julia. *Black Sun: Depression and Melancholia.* New York: Columbia University Press, 1989.
Love Peace and Poetry: Turkish Psychedelic Music. Q.D.K Media, 2005.
Pamuk, Orhan. *Istanbul: Memories of a City*. London: Faber & Faber, 2005.
—. *My Name is Red.* London: Faber & Faber, 2001.
Pasolini, Pier Paolo. *Roman Poems.* San Francisco: City Lights Books, 1986.
Richardson, Joanna, ed. *Charles Baudelaire: Selected Poems.* London: Penguin Books, 1975.
Sevda Zinçiri. Melbourne International Festival, 2005.
Schumacher, Claude, ed. *Antonin Artaud on Theatre.* London: Methuen, 1989.
Zurbrug Nicholas, ed. *Art, Performance, Media, 31 Interviews.* London and Minneapolis: University of Minnesota Press, 2004.

Websites

http://psychevanhetfolk.homestead.com/TurkishProgressive.html
http://www.onstagerecords.com/Turkish_psych/turkish_psychedelic.htm
http;//home.earthlink.net/~ikritkos/index1.html

SEPHARDIC PLACES IN SEPHARDIC SONGS

DERYA AGIŞ

Abstract

In this study, I aim to analyze some Sephardic songs, which are sung today by the Turkish Sephardic pop group Sefarad, the Jewish couple Jak and Janet Esim, and the Israeli singer Hadass Pal-Yarden. In these songs, several quarters of Istanbul, Jerusalem, and Izmir are mentioned. Istanbul and Izmir are the cities where the Jews, who had been expelled from Spain in 1492, settled down. Of course, some quarters of these cities are crucial for Sephardim – the name given to the Jews that came to the Ottoman Empire in 1492-, as they have been working and living in these places. Moreover, one cannot deny the importance of the sacred city of Jerusalem and Israel for the Sepharads. Consequently, I intend to analyze how the Sephardim mention these cities from a cognitive poetic point of view with a Gestalt-like approach to the grammatical constructions, rhymes, musical tunes, and metaphors in these songs.

Introduction: What is Cognitive Poetics?

Literary works are the products of the human mind. For this reason, the Israeli philologist Reuven Tsur proposed the theory of cognitive poetics for the interpretation of poems and poetry pieces for the first time in his book entitled *Toward a Theory of Cognitive Poetics*, published in 1992. Literary utterances are based on the SCRIPT or COGNITIVE SCHEMA that they instantiate (Tsur 232). Besides, Tsur also argues that several cognitive processes are involved in the comprehension of the semantic ambiguities. Consequently, according to Tsur (207), a theory of metaphor must satisfy four requirements of adequacy: a. to define metaphors structurally; b. to explain how human beings understand metaphorical expressions; c. to depict the relationship between this process and the process via which human beings produce and conceive literal discourse; d. to describe the relationship between these processes and the perceived effects of metaphors.

Moreover, the semantic information processing model includes a hierarchical one of 'meaning components,' 'features,' or 'semantic primitives' (Tsur 207). According to the 1992 theory of Tsur, these components are depicted via the processes of cancellation, indicated by a minus, multiplication, indicated by a plus, symbolization, abstraction, allusion, spatio-temporal continuity, richness of conceptual categories, the existences of a spiritual space, and rhymes.

In this study, I analyze the place names via some cognitive poetic tools, thus, cancellations, additions, spatio-temporal continuity, categorization of concepts, frames and spiritual spaces of songs, and the rhymes in the songs via these linguistic means. The following songs are used in this study: *Bre Sarika ('Hey, Little Sarah')*, sung by the contemporary Turkish Sephardic pop group, Sefarad, and *Ir Me Kero Madre a Yerushalayim* ('*Mother, I Want to Go to Jerusalem*'), sung by Hadass Pal-Yarden, and *Alta Alta Es la Luna* ('*The Moon Is at the Very Top'*) sung by Janet and Jak Esim.

History of the Sephardim in Turkey

In 1492, the Catholic monarchs of Spain, Isabella of Castilla, Ferdinand of Aragon and their Prime Minister Torquemada expelled the Jews from Spain, since they did not accept to convert to Catholicism (Sarhon). These Spanish Jews who migrated to North European countries did not continue to speak Judeo-Spanish, but those who settled in North Africa, and the Ottoman Empire continued to speak it; few additions and some structural readjustments were made to the language (Altabev 62 - 63).

In the Ottoman Empire, the Jews, coming from Spain, spoke the Castilian language of the Middle Ages (Shaul 12). The Jews who came from Portugal to the Ottoman Empire after their expulsion from Portugal in 1497 (Gilmer 16) spoke Portuguese. Additionally, some Jews could speak Greek, as it was the language of the Byzantines; besides, there were Venetians and Genovese, living in Turkey: they could speak Italian (Shaul 12). The lexicon of these languages together with those of Turkish, Hebrew, and Arabic entered into the Judeo-Spanish or Ladino language.

Later French words entered into the language, as French became the language of education after the establishment of l'Alliance Israelite Universelle (Shaul 13). In the successive sections, three Judeo-Spanish songs will be analyzed.

The Analysis of the song '*Bre Sarika*'

In this section, we will give the Judeo-Spanish song with the rhymes, and then pass to its translation into English.

JUDEO-SPANISH	
BRE SARIKA	
-Bre Sarika bre…	a
Trayme un poko d'agua.	b
-□Sto deskalsa	c
Ay rosiyo en basho	d
Me se yela el pye.	e
Chika i ermozika	f
-Bre Sarika bre	a
Linda i savrozika	f
Yo te va merkar	g
Sapatos d'Unkapan	h
Chizmes de Djibali	i
-No me primo a ti	j
Ke me merkes tu,	k
Tengo un padre,	l
Merkador muy grande,	m
I el me va merkar	g
Sapatos d'Unkapan,	h
Chizmes de Djibali.	i

ENGLISH
HEY LITTLE SARAH
-Hey little Sarah hey
Give me a bit of water
-I am barefooted
My foot is cold
Small and beautiful
-Hey little Sarah hey
Beautiful and sweet
I am going to buy you
Shoes from Unkapani
Boots from Cibali
-So that I buy you
Before you buy me
I have got a father,
A great seller,
And he is going to buy me
Shoes from Unkapani
Boots from Cibali

The song is about the love of a Sephardi boy for a girl called Sarah. The boy's father is a trader in the European side of Istanbul. Therefore, the boy wants to buy some shoes and boots, and give them to the girl.

Feature Cancelations and Additions

Sarah is depicted as a young and beautiful girl with the following cognitive operations of feature additions and deletions:

[+ANIMATE +HUMAN +FEMALE -OLD +BEAUTIFUL +NICE +SWEET (#SARAH)]

The singer who loves Sarah can be described with these features, as he is an animate human being; he is tired, thirsty, but generous to buy shoes:

[+ANIMATE +HUMAN +MALE +TIRED +THIRSTY +GENEROUS (#SINGER)]

Concerning the two places whose names are mentioned in the song, Unkapani is an excellent shoe market, and Cibali is a perfect boot market. These are the shopping centers of Istanbul:

[-ANIMATE –HUMAN +SHOE MARKET (#UNKAPANI)]
[-ANIMATE –HUMAN +BOOT MARKET (#CIBALI)]

Spatio - Temporal Continuity

The song begins with an imperative that attracts the attention of the listeners to the lady whom the singer wants to bring some water. The present tense is used in: -*'Sto deskalsa'* ('my feet are bare') and *'me se yela el pye'* ('my feet freeze'); and the future tense is used in '*yo te va merkar...*', ('I am going to buy you…'). The present subjunctive occurs in '…ke me merkes tu,' ('… that you buy me'), and then the present tense '*tengo un padre*' ('I have a father') appears.

The singer explains the economic condition of his family for attracting the lady.

Categorization of Concepts

The places appear as market places: Unkapani as a shoe market and Cibali as a boot market.

Frame and Spiritual Space of the Song

In the frame of the song the shoe market of Unkapani and the boot market of Cibali appear. The singer's father appears in the spiritual space of the song indicating the economic welfare of the singer.

Rhymes in the Song

Only the words '*ermozika*' ('little beautiful') and '*savrozika*' ('little sweeet') are rhyming. These indicate the beauty of the lady. These occur several times: '*bre Sarika bre*' ('hey little Sarah hey') '*sapatos d'Unkapan*,' ('shoes from Unkapani') and '*Chizmez de Djibali*' ('boots from Cibali'). These show that these places are important market places alluding to the economic welfare of the author.

The Analysis of the Song *'Ir Me Kero Madre A Yerushalayim'*

This section is in a song where a girl expresses her desire to go to the sacred city of Jerusalem.

JUDEO-SPANISH
IR ME KERO MADRE A YERUSHALAYIM
Ir me kero, madre, a Yerushalayim. a
A pizar las yervas, i artarme de ellas. b
En el me arimo yo, c
En el meafiguzyo yo, c
En el, señor de todo el mundo. d

ENGLISH
MOTHER, I WANT TO GO TO JERUSALEM
Mother, I want to go to Jerusalem.
I want to walk on the grass, till I'm sated.
I trust in Him.
I'll take refuge in Him.
In God, the ruler of the world.

Feature Cancellations and Additions

The following features depict the city of Jerusalem as a sacred place and a divine shelter:

[-ANIMATE –HUMAN +SACRED +GREEN +SHELTER OF GOD +SATIATING (#JERUSALEM)]

The description of the city of Jerusalem as a divine shelter shows that the singer is a good Jewish believer:

[+ANIMATE +HUMAN +BELIEVER (#SINGER)]

The singer mentions the sacredness of the city of Jerusalem, the greenness demonstrates that Jerusalem is a place full of hopes, and all prayers should be accepted, when one prays in Jerusalem.

Spatio-Temporal Continuity

The use of the present tense implies the strength of the desire of the singer to go to Jerusalem.

Categorization of Concepts

Jerusalem appears as a holy place. Walking on the green grass of Jerusalem and eating indicate that the singer is a strong believer. As Abraham arrived there, the city is recognized as 'the perfect city' by the Jews. The song emphasizes the holiness of Jerusalem.

Frame and Spiritual Space of the Song

In the frame of the song the singer begs her mother and implies her desire to go to Jerusalem as a good Jew.

Rhymes in the Song

The word '*yo*' ('I') appears at the ends of the verses, implying that the singer wishes to go to Jerusalem.

The Analysis of the Song '*Alta Alta Es la Luna*'

The last song we are going to analyze concerns the sorrow of a lover, who cannot reach the one s/he loves, but dreams about it. As the title of the song indicates, the moon is too far away to reach, so are the wishes of the lover.

JUDEO-SPANISH

ALTA ALTA ES LA LUNA

Alta alta es la luna a
Kuando empesa esklareser. b
Ija ermoza sin ventura c
Nunka yege a naser. b
Los ojos ya me s'incheron d
De tanto mirar l'amor e
Vapores ya van i vienen f
Letra para mi no ay. g
Letras alegres aresiviremos h
I los dos mos veremos. h
En un nido durmiremos h
I enjuntos konversaremos. h
Mama mia mi kerida i
Un suenyo me sonyi, j
En mi suenyo me sonyava k
Ke Pasha Turko me yevava k
Pasharikos chuchulean l
En los arvoles de flor. e
Ay debasho se asentan l
Los ke sufren del amor. e

ENGLISH

THE MOON IS AT THE VERY TOP

The Moon is at the very top
When it begins to shine.
Beautiful unlucky girl
Nothing is born.
My eyes are full of tears
Very much looking at the love/
Boats come and go
There is no letter for me.
We received happy letters
And the two of us we will see
In a nest we will sleep
And we will talk together.
My dear mother
I saw a dream,
In my dream I was dreaming
That the Turkish pasha was guiding me
Little birds sing
On the flower trees.
There below they sit
The ones who suffer from love.

Feature Cancellations and Additions

The fact that the moon is far away renders the girl, who cannot reach her lover, sad; however, the moon has some light, and it shines. Consequently, there is hope, as we can understand from the features assigned to the moon:

[-ANIMATE -HUMAN +HIGH +SHINING -CLOSE (#MOON)]

The moon is a far place, but it has the scope of illuminating the earth at night. Thus, every bad event may have a good reason for existence. Besides, the emotion of passionate love is extremely painful, as a person who loves another devotedly cannot keep herself / himself away from the one s/he loves:

[-ANIMATE -HUMAN +PAINFUL (#LOVE)]

Spatio - Temporal Continuity

In the first verse, the present tense is used in order to explain the present condition of the singer. In the second verse, '*incheron*' ('they got filled') is in the past tense. Besides, the sentence '*van i vienen*' ('they go and come') is in the present; a hidden hope exists in the use of the present tense after the past tense. In the third verse, the future tense is used in order to prove that there is hope.

In the fourth verse, the past tense is used in order to express the dream of the singer, and in the last verse the present tense is used in order to show that there is hope as nature demonstrates.

Categorization of Concepts

The moon is a metaphorical concept of hope that brightens the dark nights. The joy in the nature and the singing birds are the indicators of a future happiness.

Frame and Spiritual Space of the Song

In the frame of the song, the singer suffers from love. In the spiritual space of the song, the hope of the singer of receiving a response to her / his love appears, as the birds sing.

Rhymes in the Song

In the first verse, '*esklareser*' ('to shine') and '*naser*' ('to come to the world') are rhyming, indicating that the singer has the hope that s/he will see her / his lover. The light is the symbol of hope.

No rhyming words are present in the second verse. In the third verse, the following words rhyme:

…'*aresiviremos*' ('we will receive')	h
…'*veremos*' ('we will see')	h
…'*durmiremos*' ('we will sleep')	h
…'*konversaremos*' ('we will talk')	h

All verbs conjugated in the future tense rhyme. This fact alludes to the hope of the lover to see the one s/he loves.

In the fourth verse, the phrases '*sonyava*' ('s/he was dreaming') and '*yevava*' ('he was guiding') rhyme. The use of the past continuous tense shows that the lover remembers her / his dreams, and s/he wishes that they become true.

In the last verse, '*chuchulean*' ('they sing') and '*se asentan*' ('they sit') rhyme for indicating the pain of the lover, creating an opposition between the happy birds and the unhappy lover who did not receive a response to his love. Besides, the rhyming words '*flor*' ('flower') and '*amor*' ('love') imply the beauty of true love, when the two partners love each other. The flowers are naturally beautiful, and so is the true love in which one corresponds to another's love.

Conclusion

The names of the places used in these songs show that culture is an important element in the personal cognitive processes of the composers of the songs. The songs can be analyzed from a cultural – cognitive perspective, as the place names are those where a group of Sephardim lived in various parts of the world, and which are metaphorical elements for them, and so used due to socio-cultural beliefs.

Some socio-cultural beliefs play a crucial role in the use of some cognitive elements, i.e. words in these songs in the choices of place names.

The first song shows us that Unkapani and Cibali are good market places, the second song that Jerusalem is a very holy city, especially according to the Jews, and the third song that the moon is the cultural symbol of hope metaphorically, shining in the dark, and brightening the black night sky.

Works Cited

Altabev , Mary. *Judeo-Spanish in the Turkish Social Context: Language Death, Swan Song, Revival or New Arrival?* Istanbul: Isis, 2003.

Sarhon, Karen. Judeo-Spanish Language and Culture. n.d. 1 March 2006 <http://www.istanbulsephardiccenter.com/index.php?contentId=41&mid=31>.

Gilmer, Paul Gregory. *Judeo-Spanish to Turkish: Linguistic Correlates of Language Death.* PhD thesis. University of Texas - Austin, 1986.

Janet & Jak Esim. "Alta Alta Es La Luna." Mira. Istanbul, 1999.

Pal-Yarden, Hadass. "Ir Me Kero Madre a Yerushalayim." Yahudice. Istanbul, 2003.

Sefarad. "Bre Sarika." Sefarad - Volume 2. Istanbul, 2003.

Shaul, Eli. *Folklor de los Judios de Turkiya*. Istanbul: Isis, 1994.

Tsur, Reuven. *Toward a Theory of Cognitive Poetics*. Amsterdam, London, New York & Tokyo: North-Holland, 1992.

LANGUAGE AND IDENTITY IN UNITED EUROPE

VASSIL ANASTASSOV

Abstract

The subject of this paper is the linguistic background of the concept of "Pan-European Language," as a possible threat for the cultural identity of different "nations-states" – present and prospective members of the EU. It investigates the interdependence between the different languages spoken around Europe and European identity. More specifically, it deals with the fear of any specific language "pre-dominating" over the rest; a fear that stimulates the emergence of extreme ideas, ranging from a "Pan-European Pidgin" to the use of an "artificial" (Esperanto type) means of communication. It is assumed that the socio-cultural background that most EU members share predetermines the formation of a European group identity, where a "common language," as a secondary one would play a much less significant role in the process.

Introduction

The first step to make, in the search of a appropriate definition of "European Identity" is to agree on: What is Europe? A question that Michael Wintle cautiously asks himself in his impressive speech on the political associations of the Ancient Greek myth about the Phoenician princess Europa, abducted by Zeus in the outer form of a bull and Europe – the continent, throughout history. By analyzing visual images, he tries to find appropriate answers to vital questions, such as "What has been the interaction between visual images of Europe in a more conventional sense, such as European culture, European civilization, or European politics: in a word, European identity?"(5).

Wintle, like many others, who are practically and theoretically involved in the "building up" of the European Community, are often at a crossroad, as to how to define common European values in balance with

European diversity. I have deliberately chosen Wintle's parallel between European art and European identity as a starting point for my research, because language, as a human prerequisite based on symbols is not far away from art imagery, as part of the cultural identity of a group, or groups, that share neighboring territories.

In this respect Europe, inhabited by so many genetically different *ethnoi* (speaking different languages), is a place where, paradoxically enough, each separate ethnos proclaims itself European, often with disrespect for other neighbors who share the same continental territory. This is why in some Indo-European languages the words for "neighbor" are etymologically related to words, meaning "enemy, foe, impostor." This is how in the case of Europe the antagonism between "We" and the "Other" developed into dangerous nationalism, especially during the period between Napoleon and Hitler.

There can be no doubt that the awareness of many Europeans is much more national, and increasingly regional, than European [...]. But the predominance of national awareness has caused too many evils in Europe's recent history – and goes on causing them. In his emotional last speech to the European parliament, Francois Mitterand pronounced those words in which for Europe, he associated nationalism with war (Banus 158)

And, further on:

> The strength of this withdrawal nationalism stems from a very potent mixture: the atavistic fear of losing what is one's own when it comes in contact with what is someone else's [...]. Thus, the 'establishment of group identity' is often achieved 'by means of differentiation from other group identities [...], so that often "the notion of *alterity* becomes synonymous to that of *antagonism*. Internal cohesion, in the ideal nation-state model, required not only cultural, and if possible, linguistic unity, but also the referent "foreign" as a screen to reflect that cohesion [...]. This is how the concept of "cultural identity" became consecrated over a long period as a synonym of "national identity". [...] To oppose bull-headed nationalism - [...] - European awareness should be reinforced, because it means an "opening of horizons". It corresponds to the reality of cultures, where continual interchanges have produced the phenomena of blending races, changed habits and mentalities, while yet maintained identities [...] "*living together, cooperation and cultural interaction increase the possibilities and the creativity of each people, and they do not cause changes to any culture*, let alone threaten any language, nationality or culture with extinction"[...]. It is *evident, that where there are contacts, dialogue, interchange, there can be change. But not every change means risking the loss of identity. Because 'identity' is a more flexible and multi-level entity,*

> *and not a monolithic one in which different elements are unable to co-exist.* (ibid, italics added)

Consequently:

> Europe especially has been living for many centuries in continual interchange, in a continuous interaction between the cultures – in assuming elements that have become common without, thereby, losing diversity. [...] and the presence of characters and stories from different national sources in the cultural life of so many citizens has presented no problem for plurality, rather it has been undoubtedly 'an enriching factor and a distinguishing mark of the people's Europe (ibid.).

Cultural Environment and Group Identity

Common cultural heritage (commonly shared cultural constructs) shapes up cultural identity in its way to group identity. How is language part of the cultural identity of a certain group, then?

The role of language in society and politics often holds a secondary position in modern political theory. Linguistics as a separate subject has not been so far seriously considered as a reliable source for the establishment of political science as a science. In fact, language analysis can contribute to the successful investigation of unique or specific languages and it can, thus, support the idea that national language (with its assumed significance in the process of nation building) should be reconsidered in the context of United Europe - building. I should agree here with Harold Schiffman who claims that "purely" linguistic and language-policy analytical aims often overlap in the field of semantics where the cultural load turns language code into real language.

[...] if language policy is not deeply rooted in the linguistic culture of a language group, it is not going to fit the needs of its speakers as well. Language policy is therefore not just a text, a sentence or two in the legal code, it is a belief system, a collection of ideas and decisions and attitudes about language. It is of course a cultural construct, but it is either in tune with the values of the linguistic culture or it is in serious trouble [...]. (Schiffman 56)

Therefore, language policy should be able to define the linguistic culture of a language group. Using Schiffman's idea of language policy as a belief system in the context of the nation-state we can easily understand the conventional idea of language as the landmark of national identity as an assumption that "there exists somewhere, perhaps in the past, or in a particular textual tradition a state of 'purity' that the language can aspire to return to" (ibid).

This assumption, by itself, can easily lead to a "una lingua – una patria" type of language nationalism.

Given the history of any existing language on earth, there is no evidence for an entirely authentic ethnic language. A possible attempt to define what makes a language really "unique" would reveal that there are no strict boundaries between languages, let alone any correspondence between languages and nation-state borders. This is where the attempts of language purists to "excavate an absolutely pure" stage in the history of any language definitely fail. The "uniqueness" of any language is not endangered - on the contrary - it could even be enriched by the penetration of "alien" language elements, which become an inseparable part of its own system. In the case of, say, bilingual societies, the different languages which speech communities use can exist there without any damage to their specific character. When languages are in contact, understanding is possible when the two speech communities have too much in common in terms of cultural interaction: traditions, life-style etc. so that for a longer period of time they get to know each others "language building bricks." In other words,

> If we want to understand the role of languages in people's lives we must go beyond the study of their grammar and venture into the world of social action, where words are embedded in and constitutive of specific cultural activities such as telling a story asking for a favor, greeting, showing respect, praying, giving directions, reading, insulting, praising arguing in court, making a toast, or explaining a political agenda. (Duranti 1)

We may then assume then that in terms of language "specificity" there are no absolutely pure ethnic languages because there are no genetically absolutely pur *ethnoi.* Since, as we have pointed out before, specific languages have no strict borders, it will be easy to infer that borders between separate languages themselves would also blur. Furthermore, it would be enough to fix these borders if they were given once for ever. The big problem here is that they constantly change along with other changes in the structure of the different *ethnoi.*

Language innovation occurs in the cultural deixis of *here* and *now.* That is why it is the legitimate, implicit human (i.e., non-linguistic) knowledge about language. Homo Loquens locates himself then in space by means of a temporal dimension: "now" goes hand in hand with "here" whilst "then" is associated with "there."

It will be easy to infer that in terms of space "he" will be located in an environment where "he" will share the same type of cultural experience with the rest of the other representatives of the same species and react to it with the same type of communicative creativity. The implication is that in an environment where our "talking man" can express the same type of feelings and emotions as the rest of the co-inhabitants, it is not the linguistic "authenticity" or "purity" which makes interaction possible but the common background which ties up the population of a certain area into a society with its own cultural parameters. So, on an interethnic scale – the more the inhabitants share equivalent cultural values, the more they tend do develop a real group identity, even though not necessarily associated with group language.

Common European Language, Different European Languages, European Identity: Is the Idea of a Common European Language Relevant to the European Group Identity Formation Process?

Europe has been for a long time sharing common cultural constructs with no common language as a must. There are plenty of examples from literature, music, painting and architecture:

Without the Italian, Petrarch, or the Frenchman, Ronsard, for example, it is impossible to imagine Shakespeare's sonnets; without a theological debate of Spanish Baroque it would be impossible to imagine a European phenomenon like Mozrat's Don Giovanni; [...] without the Schlegel brothers in Germany [...] Spain would have never rediscovered its own Baroque drama; and a long etcetera that can turn the hero of an eighth century song into the protagonist – of one of the twentieth century novels said to be decisive for the history of literature (James Joyce's Ulysses) [...] While, during the Baroque era, Italian architects not only built churches in Prague, Vienna or Madrid, they also prepared the scenery for plays in the Spanish Court of Aranjuez; a Greek painter sojourned in Italy and set himself in Spain (Domenicos Theotocopuli, better known as El Greco); Goethe's work would be unthinkable without his trip to Italy; from his native Hamburg, Brahms moved to Vienna, where he discovered the Hungarian world; and the Viennese, Mahler, in the third movement of his First symphony, uses the theme of a French children's song ("Frère Jacques, Frère Jacques, dormez-vous?") [...] (Banús).

Throughout human history there have been lots of cases where societies were shaped upon the basis of different ethnic strata with their different linguo-cultural properties. What makes the specific "socium" is not, therefore, the commonly shared language but the commonly shared cultural experience. In a "United Europe," where the population is increasingly sharing a similar cultural environment, the "language-identity" criterium gradually loses its relevance and gives way to some kind of "Lingua Franca" providing understanding.

A European "Lingua Franca?"

Lingua Franca is a term that was coined up to denote a kind of an interethnic language motivated by the needs of trade and cultural interaction around the region. As the traditional definition goes, "a mixture of French, Arabic and Spanish in the Levant. [...] any language, serving as a medium between different nations whose languages are not the same system, providing understanding" (Anastassov 259).

Today, most European interaction that requires a common language is maintained in English (although French and German are also largely used) Frequently appearing opinions determine English and its growing influence as the result of the political, economical and military power of (mainly) the USA. Many claim that it negatively affects the cultural independence of the separate national languages by penetrating alien elements with alien cultural connotation into their language systems. As for the needs of multi-lingual Europe, there are many who would argue that English is not the most appropriate means of "pan-European" communication. The French show signs of nostalgic reference back to the times when it used to be much more in the world of diplomacy. The politically neutral character of Esperanto seems to be a popular option for many people. They tend to underestimate the fact it failed to become the "common means of language interaction." The reason is that there are no socio-cultural criteria to integrate "Esperanto native speakers" into a Esperanto "group identity."

The use of such a common means of language interaction is often based on political factors. But even if there is an initial political reason for the language dominance, it often loses it over time and starts functioning as the symbol of socio-cultural integrity of a certain society. For example, "Athenian political dominance lasted less than a century, but the prestige of Athenian literature and of Athenian speech remained, and from it developed the 'koinē'. This word means 'shared, common, popular', and was indeed the common language of a large area for something like a thousand years" (Barber 56).

The Latin of the Roman Empire used to function as a common means of cultural interaction long after the land of Julius Caesar and Nero collapsed. The Old Church Slavic was used as the official state language of Mediaeval Valachia. The Spanish language of Latin America has nothing of the political connotation of the Spanish language of the fifteen-century Conquista. The Portunhol Portuguese of Brazil is even phonetically far away from the language of Imperial Portugal. Modern Greek was in its time used as a "Lingua Franca" in the regions of Istanbul and West Anatolia in Ottoman Turkey. The Russian language of the post-USSR era in the Turkic republics of Central Asia now plays the same role that Arabic used to play at the times of semi-legendary Turkestan. And, of course, the English of the British Empire and (American) English today are good examples of how the language of predominant power can lose its political connotation.

The conclusion is that the "Lingua Franca" type of languages appear on the historical stage *a posteriori* possible political hegemony in order to fulfill the necessity of cultural interaction around an area/territory with no common language. In the case of United Europe, where all the efforts since the very beginning of the building process have been dedicated to avoiding power misbalance, the issue of a prospective hegemonic language is absolutely irrelevant.

Conclusion

People's desire for language-culture interaction could be traced as far back as the legend of the Tower of Babel. If we refer back to the previously discussed issue of language as part of the cultural creativity of Man, the Biblical metaphor would simply reveal that in our modern world – or more specifically in modern Europe - it is the commonly shared feelings emotions, needs, etc. and not the common language that provide the creative cultural background for common understanding.

In our attempts to identify the cultural paradigm of United Europe, in the way Wintle analyzed it, we should always remember that was not on the territory of today's Europe where Zeus abducted the princess. "European Culture" or "European Civilization" belongs to all of us. It is the best possible way for the contemporary Homo Loquens to accept diversity within our community in the globalized world.

Works Cited

Anastassov, Vassil. "The Lingua Franca of the Past and Today: Political Dominance or Cultural Need?" *Culture in International Relations.* Ed.Zuzana Lehmannova. Prague, 2005. 259-264.

Banus, Enrique. "Cultural Policy in the EU and the European Identity." *European Integration in the 21st Century: Unity in Diversity?* Eds. M. Farrell, S. Fella and M. Newman. London: Sage. 158-83.

Barber, Charles. *The English Language. A Historical Introduction.* Cambridge: Cambridge UP , 2000.

Duranti, Alessandro. *Linguistic Anthropology: History, Ideas, and Issues in Linguistic Anthropology*. Cambridge, MA: Blackwell, 2001.

Schiffman, H. *Linguistic Culture and Language Policy*. London: Routledge, 1998.

Wintle, Michael. *Europa and the Bull, Europe,and European Studies.Visual Images as Historical Source Material.* Amsterdam: Vossiuspers UvA, 2004.

LANGUAGE POLICY OF THE EUROPEAN UNION

PAVEL SENDERÁK

Abstract

The Language Policy of the European Union (LPEU) is closely related to the Union's cultural policy, political integration and, implicitly, to its economic integration. It consists of a set of diffuse EU regulations, decisions and resolutions. The policy is rather incoherent and some of its underlying principles are contradictory. Its main principles include multilingualism and an equal treatment of languages enjoying the status of official and working languages. At the same time, it is expected to enable and/or facilitate an effective communication among citizens and EU institutions. It is possible to identify several problems in its implementation, caused by its inherent contradictions and financial and qualitative costs. The LPEU has never been actually implemented. We consider its present configuration unsustainable for the future and outline various suggestions for its modification. Finally, we present our own option of its optimisation.

European Integration and its Language Dimension

One of the main driving forces of European integration has been the overcoming of traditional national contentions between states and nations. This should take place without threatening the historically formed cultural identities, at least those corresponding to member-states' identities (Kraus, "Between Mill and Hallstein" 300). In 2003, the Convention on the Future of Europe decided that the new motto of the enlarged EU shall become "United in diversity." Such a motto proved increasing supranationalist trends, while, paradoxically, reasserted that cultural identities of member states cannot be threatened (Toggenburg 6). However, the endeavour after political integration through attempts to form or to "reveal" the common European cultural heritage, European identity, *demos* and the public sphere is challenging these cultural identities.

The European Community (EC)/European Union (EU) has always been exclusively constituted by states and in most of them, language has become one of the main components of national identity. The language issue is for some of them such a delicate topic that the possible integration in this field has become a political *non dit*. Since the establishment of the European Economic Community (EEC), there has always been an absolute dominance of a purely intergovernmental approach to the linguistic regime, which has not altered. The European Coal and Steel Community (ECSC) and the EEC have agreed on multilingualism as a linguistic regime of their operation. As the regime applies only to member states' official languages, main actors are not national or linguistic communities, but states. The extent to which the declared cultural pluralism is only a concession to member states remains a question.

Official setting and the legal basis of the EU Language Policy

Most of the European countries are principally culturally homogenous, but the continent as such is highly diverse.[1] In the EU, the plurality principle is embodied in institutional multilingualism, but the LPEU also impacts on the various educational and cultural policies and the EU's attitudes towards regional and minority languages (Caviedes 255).

The ECSC, established in 1952, consisted of six members and had four official and working languages.[2] The Rome Treaty of 1957 states in Article 217 that the language regime of institutions will be set by unanimous decision of the Council. Although the Treaty itself did not invoke a concrete language plan, the multilingualism principle was reflected in its form, as its original was composed in four languages and all

---

[1] "Here, a central aspect is that this homogeneity is not a 'natural' outcome of a quasi-evolutionary process, but rather reflects the deliberate use of political power by majority elites" (Kraus 671, 2003). The internal linguistic constellation of each state is distinctive. Linguistic configuration of the EU consists of: official state languages of member states, most of which enjoy the status of its official and working languages, several regional languages with local official status (such as Basque or Catalan), regional languages without such status (such as Corsican), non-territorial languages (such as Romani), and non-territorial languages of immigrant communities (such as Turkish, Arabic, Chinese or Hindu) (Mamadouh 134).

[2] This means that all EU legislation must be published in all the member states' official languages, because it also affects the national law. All institutions should be able to communicate in all official languages (Caviedes 252).

the language versions were stated to be equally authentic (Liddicoat 22). Council Regulation 1/1958 stipulated that the EEC shall have four official and working languages. Thus, the parallelism of official and working languages was established. The official language also automatically acquires a working language status. The Regulation also implemented the principle of an absolute equality of these languages. Both these rules have their origin in the Treaty Establishing the ECSC. Their validity has not been discontinued nor challenged by any of the subsequent treaties. EU language pluralism cannot be understood as an absolute principle, as it, with several exceptions, applies only to official languages of member states (ibid). In fact, the EEC language plan had ratified language plans of member countries at the supranational level.

According to the Council Regulation 1/1958, "the institutions of the Community may stipulate in their rules of procedure which of the languages are to be used in specific cases. The languages to be used in the proceedings of the Court of Justice shall be laid down in its rules of procedure. If a Member State has more than one official language, the language to be used shall, at the request of such State, be governed by the general rules of its law." Such a linguistic pre-set has left considerable space for modifications by the EEC institutions. Gradual enlargement of the EC/EU has always brought an increase in the number of official and working languages. The common practice has become that the member state nominates a language, which will become an official and working language of the EC/EU (Liddicoat 36). Two exceptions to this practice occurred in cases of Irish and Luxembourgish.[3] As the EU is a descendent of the EEC/EC, it also reassumed its language regime by Article A of the Maastricht Treaty. After the 2007 enlargement, the EU has 23 official and working languages.

Although largely neglected, the LPEU is closely associated with the concept of EU citizenship, because it sets up several linguistic rights. Besides, several language rights arise from two basic EU social rights – free movement of individuals and the right of EU citizens to settle in any member state. Moreover, the Charter of Fundamental Rights of the European Union of 2000 prohibits discrimination based on language.

[3] Irish (Gaelic) has not become an official and working language of the EC after the accession of Ireland in 1972. For EC/EU purposes it had the status of "language of treaties." Paradoxically, the Irish constitution recognizes Irish as the national language while English is considered the second official language (Alcaraz 9). Luxembourgish became the official language in Luxembourg in 1983, but it has never been granted the status of EC/EU's official and working language.

However, this prohibition fully applies only to speakers of the (official) language of another member state (Liddicoat 30-31).

Practical implementation of the LPEU

Peter A. Kraus argues that "the guiding parameters of European language policy have remained rather diffuse, and the basic procedures designed to regulate linguistic pluralism are not free from contradictions" (Kraus, "Cultural Pluralism" 675). Indeed, full multilingualism formally exists, but it has never been implemented in reality. The number of languages actually used is limited in some areas and aspects (Liddicoat 22). We may distinguish at least four domains of the LPEU's practical implementation: 1) *language regimes of institutions*; 2) *translation and interpretation services*; 3) *language-in-education planning* and 4) *other languages – minority, regional and immigrant languages*.

Institutional multilingualism is based on the Council Regulation 1/1958. "Because the regulation is rather equivocal, there is some variety between the linguistic regimes of the different EU institutions and bodies" (Mamadouh 134). Internal provisions of the main EU institutions determine that all official and working languages are equal, but theory and practice differ. "In general terms, the situation can be described as a regime of formal institutional multilingualism with the use of two languages in more informal settings: English and French" (Ibid).

In the Council, full multilingualism is applied only at formal meetings. In the Coreper, based on the consensus of all parties, English, French or German is used. Generally speaking, these three languages prevail over the rest in Council's working. The Commission has to publish its White and Green Papers, laws, press releases, Official Journal and leaflets for the public in all official languages (Muller 42). In its daily operation, only French and English are being utilized within the Commission. In the EP, multilingualism is being adhered to at all meetings and for all documents. Its Rules of Procedure also constitute and exception, because it is in certain cases allowing use of other languages. The European Court of Justice (ECJ) has a specific language regime not fully compatible with the rest of the EU. Its main working language is French, together with the language chosen at the beginning of legal proceedings (Mamadouh 134). EU bodies and agencies are using only a limited number of languages and the number and choice of languages vary among them. Language regimes are diverse because general rules of choosing working language(s) have never been clearly determined (Muller 45). Thus, in the course of time, a group of privileged languages has emerged within EU institutions.

English, French and, to a lesser extent, German have become *de facto* the only working languages of informal meetings and internal communication.

The structure of providing translation and interpretation services is very complex. Interpretation is carried out by three separate offices and translations are provided by nine. In 2000, there were 3250 translators and 850 interpreters working for the EU. These figures did not include freelance translators and interpreters, providing approximately 22% of these services as well as supporting administrative staff (Gazzola). In 2000, the Commission announced that costs of language services for 1999 amounted to €685.9 million, approximately €2 per citizen, 0.8% of EU budget and 15% of all administrative expenses. Michelle Gazzola points out that these figures are incomplete as some costs are excluded and as the EU tends to undervalue these expenditures (Ibid)[4]. Nicolas-Jean Brehon argues that annual costs of language services before 1999 amounted to €1.7 billion, approximately 2% of the budget (Brehon). Nevertheless, all these figures are referring only to the EU-15, with 11 languages and 110 one-way combinations between them.

The main objective of the language-in-education planning, as articulated by the EU, is that citizens learn three official EU languages (European Commission). This reflects the agenda of political integration, development of democratic citizenship (Mamadouh 134) and creation of European identity.

As long as the competence of recognition of language status is exclusively in hands of member states, EU institutions lack a direct authority with regard to regional and minority languages (Alcaraz 13). An important milestone was the Council's 2005 decision to allow official use of other languages within the Council and other institutions and bodies of the EU. "These conclusions relate to languages other than the languages referred to in Council Regulation No 1/1958 whose status is recognised by the Constitution of a Member State on all or part of its territory or the use of which as a national language is authorised by law" (SCICNEWS n.pag.). Within the LPEU, the position of immigrant languages of the European origin is rather poor. Non-European immigrant languages are mostly ignored and are explicitly excluded from the provisions of the European Charter for Regional and Minority Languages.

---

[4] Emily van Someren argued that these costs in 2003 amounted to €2.55 per citizen, while the total EU expenses on education, justice and home affairs, press and communication, and health and consumer protection represented only €2.36 per citizen (van Someren 20). M. Gazzola expected that after the 2004 enlargement, expenses on language services will increase to €3 per citizen (Gazzola).

Problems Arising from the Present LPEU and its Possible Modification

We may identify at least five problematic aspects of the present LPEU. First is the *exclusivity of granting the status of official and working language of the EU*, as it may be granted only to an official language of the member state or one of them in case of a multilingual state. Therefore, it is the individual Member States who decide whether a language is eligible to become the EU official and working language. The position of unofficial languages in the EU depends on their position in respective member states. Thanks to efforts of Ireland and Spain, Irish (Gaelic), previously considered as the "language of treaties", had become an official language in 2007, and Spanish regional languages (Basque, Catalan/Valencian and Galician) have *de facto* gained the status of "language of treaties" (Muller 51). This contrasts with the approach of France, which according to its Constitution, does not officially recognize any regional, minority, or any other type of language but French within its territory.

The LPEU seems inconsistent. While Maltese with around 380 000 speakers (Yves n.pag.) has the status of official and working language, larger regional and minority language groups (such as Catalan, Basque, Galician, Frisian or Welsh) are deprived of all rights arising from such a status. This also applies to languages without the regional or minority language status (Caviedes 258). Languages whose speakers are not living concentrated in one region but are dispersed across the whole territory or in several states, such as numerous immigrant languages or Romani, seem to have the worst position of all (Ibid. 260-61). Nevertheless, the 2005 extension of the status of Spanish regional languages should be considered a step forward. We may conclude that linguistic diversity as a basic principle of the LPEU is *de facto* exclusive and does not take into account non-traditional and non-European languages.

The second problem is the *LPEU's impracticability and the actual inequality of languages with the official and working language status*. The status should guarantee absolute equality of these languages and their speakers. Instead, EU institutions and bodies are limiting the number of working languages and also, though rarely, the official ones. As the difference between official and working languages has never been formally fixed, it seems that they fail to implement the principle of their absolute equality.

While full equality could have been executable with four languages of the founding Six, although even at that time French dominated (Muller

47), full equality of more than 20 languages would paralyze institutions and cause a substantial increase in expenses. This equality is thus only formal: "Some languages are more equal than others" (Loos 44). In fact, an unofficial hierarchy of languages emerged within the EU's internal structure. Three languages have a dominant position. We may see an ever-strengthening position of English, stagnating or slightly weakening position of French, and by far weaker, but rather stable position of German (Loos 47, Quell 57-76 and Muller 48). This may be observed in the overview of the actual number of language versions of main websites of EU institutions, bodies and agencies listed below. Out of a total 34, only three websites provide all 23 language mutations. All of them are accessible in English only, 19 in French, and 16 in German.

Figure 1: Language versions of main websites of EU institutions, bodies and agencies as of 5 May 2007

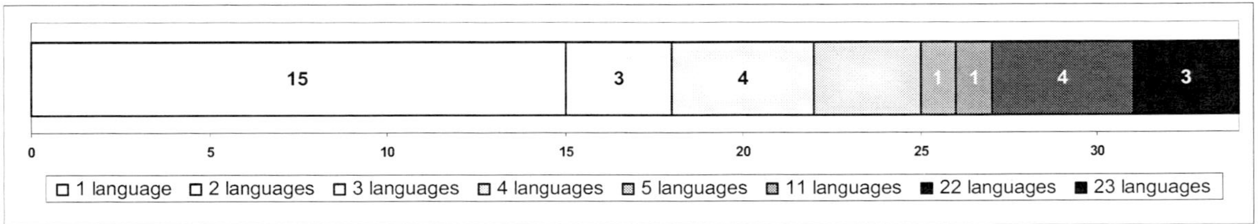

Source: Author.

Figure 2: Language versions of main websites of EU institutions, bodies and agencies as of 5 May 2007

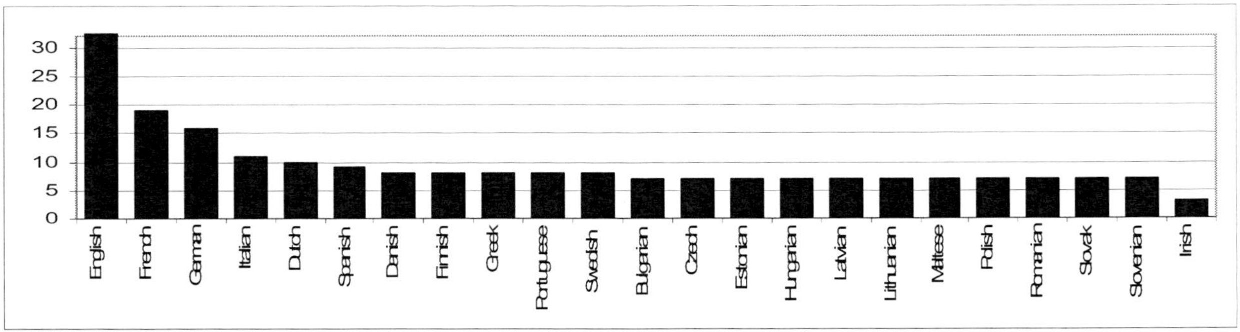

Source: Author.

It appears that controversies among the member states over the LPEU are caused by the way Council Regulation 1/1958 is interpreted. As long as the position of their own language is preserved, states are accepting its reading in the sense that although all official and working languages are formally equal, it is necessary to reduce the number of working languages to enable an effective operation of the EU and its institutions. Should their language be displaced from the list of interpreted and translated languages, they protest and make reference to the principle of absolute equality. The main problem of the first interpretation is the actual choice of languages to be preserved. We may consider many variables (total number of speakers in the EU, number of member states in which it is an official language, worldwide significance etc.), but only subjectively assess their relative importance.

Costs of multilingualism. There are two types of costs arising from the EU's multilingual operation: 1) financial costs and 2) costs caused by inaccuracies, errors and other imperfections of translation and interpretation services that cannot be expressed in numbers. In both these respects, every enlargement brings an apparent burden. While there were 12 combinations with 4 languages, it was 110 with 11 languages, 380 with 20 languages, and with 23 languages of the EU-27, the number of combinations grew to 506.[5]

Multilingualism as an obstacle to further economic and political integration. The main shortcoming of the LPEU is its over-emphasizing of the cultural aspect of language (Kraus, "Cultural Pluralism" 675). For the sake of effective economic and political integration it is necessary to alter this understanding of language and see it as a means of communication. If the building of "political Europe" is conditioned by the principle that none of present cultures should, at least formally, prevail, the EU is trying to establish *political citizenship*. Existence of a public sphere as a discussion forum for all concerned is critical for its emergence. However, its formation is being complicated by linguistic pluralism as barriers to functional communication impede the formation of a common political identity (Kraus, "Between Mill and Hallstein" 308). Despite the fact that the EEC limited its language policy only to its internal operation, the LPEU currently also has an effect on language use beyond the EU structure. A common means of communication or communication as easy

[5] The number of combinations between individual languages increases exponentially, because, according to the most common practice, it is necessary that the interpreter and translator convert writing or speech into their mother language. The number of combinations it then counted as $N = L^2 - L$, where N is the total number of combinations and L the number of languages.

and comprehensible as possible are important for political integration (Alcaraz 5). Present multilingualism – protecting member states' cultural identities – also hampers the economic integration as it is obstructing two EU freedoms – free movement of labour and individuals. While the mobility of labour is essential for the single market, language barriers are the main cause of a low number of citizens working abroad (Liddicoat and Muller 3). Karis Muller argues that it may be possible to challenge the linguistic legislation by arguing that measures for protection of languages are principally in conflict with four freedoms introduced by the Single European Act (Muller 49). Protection of member states' cultural identities and the EU's inability to intervene in internal language arrangements of its member states may lead to ECJ's inability to provide an unequivocal ruling.[6]

Vagueness and non-conceptuality of the LPEU. The LPEU lacks a coherent conception and is formulated unclearly and vaguely. It has no specific objective and, with an exception of Regulation 1/1958, consists of a rambling set of documents addressing it only marginally (Muller 41). It cannot be considered an active EU policy. While the EU supports multilingualism and linguistic diversity, it is generally accepted that a limited language regime is necessary for a proper functioning of institutions. Language choice is a highly politicized and controversial matter that the EU strives to avoid, which in turn leads to an absence of any active language policy (Loos 37). The LPEU's vagueness is caused by the cautiousness it is dealt with. Since the LPEU does not practically endanger cultural identities of member states and maintains the dominant position of their official languages, it has not become a political matter. With some exceptions, the debate about the rationalization of the linguistic operation of the EU is a political taboo.

It seems that language and cultural diversity are understood in the same manner – instead of an actual appreciation of various cultures and languages, the EU attempts to find noble terms for conservation of the present seemingly havenless condition, where none of its policies may endanger member states' cultural identities, particularly of stronger ones.

Abram de Swaan claims that the "multiplicity of languages actually subverts diversity: the more languages that compete, the more *English* will take hold. In the general confusion of tongue, in which no indigenous language can predominate, English automatically imposes itself as the sole, obvious, solution" (de Swaan n.pag.). Although the insistence on a

---

[6] Authors often point to two cases: Groener vs. Minister for Education and Salomone Haim vs. Kassenzahnärztliche Vereinigung Nordrhein (Aziz 286, Aziz and Parijs 14).

present language regime supposedly conserves institutional multilingualism, it *de facto* favors language homogenization, because of the growing use of English in informal settings (Loos 47). By doing nothing, the Commission helps to strengthen the position of English as the only way out of the confusion of tongues (de Swaan n.pag.). At present, English is the main means of communication in the EU, and thus, it seems that the problem is being solved by itself.

Possible modifications of the LPEU

The EU is not, despite the pressure of several national delegations, able to effectively hinder the growing dominance of English, caused in large part by factors out of reach of her policies (Olsen 19-20). Multilingualism is being implemented insufficiently in its width, with regard to language democracy and diversity, as well as in its depth, with regard to communication and effective operation. Change of the LPEU seems to be inevitable, but it will be a delicate and complicated process (Cajvaneanu 22). We may divide all possible modifications into two groups according to whether the parallelism of official and working languages shall be discontinued or not[7] and then regarding their impact on the number of languages with such a status into three groups: 1) single official/working language; 2) reduction of official/working languages and 3) increase in the number of official/working languages.

A *single language* could enable fast and clear communication with relatively low costs, increase efficiency of the decision-making process (Van Parijs 111-152; Mateja n.pag.), and stimulate the formation of European identity (Muller 51), as well as the emergence of the European public sphere. One of the present official languages or the most neutral language may become the *lingua franca*. We assume that the only present official language which could aspire to such a position is English. Some propose French, but its position is far less strong than the position of English and criticisms against it would probably be even more acute. Arguments supporting the choice of English include that English is the main international language (Loos 46), is largely de-nationalized and de-territorialized (Muller 46-47), may be seen as the EU's *lingua franca de facto*. Almost 90% of students study English as a second language (de Swaan), and, finally, 56% of EU citizens claim to speak English (Cajvaneanu 22). As for artificial languages (Latin, Esperanto and

[7] Karis Muller argues that the strengthening of English is caused precisely by the parallelism, since democracy is impracticable with so many languages (Muller 54).

Europanto), the main benefit would be that these languages seem to be more neutral than others, but it is likely that problems with their instability and variability in time and place would arise (Černý 28-31).

Reduction of the number of official/working languages. Even though there is a considerably higher chance that the LPEU could be fully implemented, this option seems unfeasible, because of the almost certain deadlock as regards the choice of few "privileged" languages with consent of all the member states over such a choice.

The option of *increasing the number of official/working languages* appears to be the most democratic one (Mamadouh 135), but the problem will be again the choice of such languages or formulation of clear and transparent criteria for extending the status to a particular language. A drastic increase of expenses on language services should be anticipated and in case all of them would also be working languages, LPEU's impracticability is almost certain.

Proposed Modification of the LPEU

We attempt to identify the best possible LPEU modification based on five problematic areas elaborated above. Choosing a single language – *lingua franca* – seems to be the best available option. This language shall become part of the language repertoire of all EU citizens as one, or the only, foreign language he or she possesses. Single language represents a suitable, but surely not the only, solution of three problematic areas. Prospects of full practicability are considerably high. We may also expect a significant decrease in costs of both financial and qualitative character. This option would also eliminate the problem of an unequal position and treatment of languages with the status of official and working language. This would ensure sufficient transparency, but the criteria of a democratic choice of language would be highly dependent on which language should have this position, as it should be as neutral as possible.

We assume that the most suitable language to become the EU *lingua franca* is English. The problem with selecting English is that it is not a neutral language and its native speakers would have an advantage over others. However, this criticism of selecting English is being mitigated by two factors. Firstly, current LPEU permits English to have such a position even today, but as long as there is not a formal recognition, compensation cannot be demanded. The second factor is presented by Phillip Van Parijs – the knowledge of English as a foreign language is as widespread and evenly distributed insofar, the adoption of English as a *lingua franca* does not in the interview or the competition represent a substantial advantage for native speakers or for non-native speakers with a privileged access to the language.

It is very unlikely that EU member states would easily accept the exclusion of their language from the list of official languages. The formal discontinuation of the parallelism of official and working languages and their distinction may offer a possible path forward. While the status of official language might, as an important symbol, become more flexible and thus could also be granted to a wider range of languages than today, there would be solely one working language – English. By putting the parallelism of official and working languages to its end, we would abstractly distinguish between understanding language as part of cultural and national identity and language as a means of communication.

The costs of institutional multilingualism are only partly financial. "The costs of reduced internal worker mobility and of net outflows of human capital are bound to become far more significant as both EU-wide economic integration and economic globalisation proceed" (Aziz and Parijs 16). A single language would stimulate the mobility of labour force and people and thus help enhance economic integration. Citizens would not be required to invest in learning more than one language – English – and would be able to communicate and work in the whole EU. An important feature of such a modification should be that all relevant authorities of EU member states should be able to communicate with citizens in *lingua franca*. This would benefit both EU citizens without knowledge of the host country's language as well as members of minority, regional and immigrant language communities as they would not be required to learn the language of the country of their residence.

Summed up, our suggestion is the formal distinction between official and working language status and the obligation of relevant authorities of member states and the EU to communicate with citizens in a *lingua franca*. A single language may stimulate the formation of a European political identity that would not threaten national, ethnic or regional identities. While the plurality of official languages would still symbolize the EU's cultural diversity, the language harmonization would simplify or enable communication across national and language boundaries. Yet, such language integration would not necessarily evoke the EU's cultural homogenization.

The legitimization of the EU's single language would be a long and slow process. On the other hand, despite numerous efforts, the process of globalization imposes further language and cultural homogenization every day. It appears that this scenario is hard to avoid. If the EU is able to adequately react to globalizing trends, it may, by the introduction of a common language of EU citizens, facilitate the preservation of cultural and language diversity by removing the pressure on it.

Works Cited

Alcaraz, Manuel. "Language and Institutions in the European Union." *Mercator Working Papers* 5 (Dec. 2001): 1-14. 14 Jan. 2008 <http://www.ciemen.org/mercator/pdf/wp5-def-ang.PDF>

Aziz, Miriam. "Mainstreaming the Duty of Clarity and Transparency as Part of Good Administrative Practice in the EU." *European Law Journal* 10.3 (2004): 282-295.

Aziz, Miriam and Phillippe Van Parijs. "Linguistic Legislation for XXIst Century Europe." *Université Catholique de Louvain.* Jun. 2002. 14 Jan. 2008 <www.etes.ucl.ac.be/DOCH/DOCH%2087.pdf>

Brehon, Nicolas-Jean. "Le coût du multilinguisme." *Le monde* 30 Nov. 1999. 14 Jan 2008 <http://www.finances-europe.com/article.php?id=8>

Cajvaneanu, Doina. "Regulating the New Babylon." *The Prague Journal of Central European Affairs* 4.2 (2002): 22-23.

Caviedes, Alexander. "The Role of Language in Nation-Building within the European Union." *Dialectical Anthropology* 27 (2003): 249-268.

Černý, Jiří. *Dějiny lingvistiky.* Olomouc: Votobia, 1996.

De Swaan, Abram. "Endangered Languages, sociolinguistics, and linguistic sentimentalism." *European Review* 12.4 (Oct. 2004): 567-580. 14 Jan. 2008 <http://www.deswaan.com/engels/uk_index.htm>

European Commision. *White paper on education and training: teaching and learning: towards the learning society. European Commission.* 1995. 14 Jan. 2008 <http://europa.eu.int/comm/education/doc/official/keydoc/lb-en.pdf>

Gazzola, Michele. "Multilingualism In The European Union And Its Costs." *European Commision – Representation in Italy*. 22 Feb. 2006. 14 Jan. 2008 <http://www.uemilano.it/multilinguismo/multilinguismo_unione_europea_en.htm>

"Irish, and Spanish regional languages." *SCICNEWS* (Newsletter 99) 27 July 2005. 14 Jan. 2008 <http://scic.cec.eu.int/scicnews/2005/050727/news_03.htm>

Kraus, Peter A. "Cultural Pluralism and European Polity-Building: Neither Westphalia nor Cosmopolis." *Journal of Common Market Studies* 41.4 (2003): 665-686.

Kraus, Peter A. "Between Mill and Hallstein. Cultural diversity as a challenge to European integration." *Cultural Diversity versus Economic Solidarity. Proceedings of the Seventh Francqui Colloquium.* Ed. Phillippe Van Parijs. Brussels: De Boeck, 2004. 299-314.

Liddicoat, Anthony J. "Language planning, linguistic diversity and democracy in Europe." *Perspectives on Europe: language issues and language planning in Europe.* Eds. Anthony J. Liddicoat and Karis Muller. Melbourne: Language Australia Ltd., 2002. 21-39.

Liddicoat, Anthony J. and Karis Muller. "Language issues and language planning in Europe." *Perspectives on Europe: language issues and language planning in Europe.* Eds. Anthony J. Liddicoat and Karis Muller. Melbourne: Language Australia Ltd., 2002. 1-9.

Loos Eugène. "Language Choice, Lingustic Capital and Symbolic Domination in the European Union." *Language Problems & Language Planning* 24.1 (2000): 37-53.

Mamadouh, Virginie. "Beyond nationalism: Three visions of the European Union and their implications for the linguistic regime of its institutions." *GeoJournal* 48.2 (1999): 133-144.

Mateja, Peter. "EU Working Language." *IDEA*. 24 Jul. 2006. 14 Jan. 2008 <http://www.idebate.org/debatabase/topic_details.php?topicID=288>

Muller, Karis. "Language competition in European Community institutions." *Perspectives on Europe: language issues and language planning in Europe.* Eds. Anthony J. Liddicoat and Karis Muller. Melbourne: Language Australia Ltd., 2002. 41-60.

Olsen, Johan P. "The political organization of Europe: Differentiation and unification." *ARENA Working Paper Series* 23 (23 Sep. 2005): 1-38. 14 Jan. 2008 <http://www.arena.uio.no/publications/working-papers2005/papers/wp05_23.pdf>

Quell, Carsten. "Language Choice in Multilingual Institutions: A Case Study at the European Commision with Particular Reference to the Role of English, French and German as Working Languages." *Multilingua* 16.1 (1997): 57-76.

Toggenburg, Gabriel N. "'United in diversity': Some thoughts on the new motto of the enlarged Union." Paper presented at II Mercator International Symposium: Europe 2004: A new framework for all languages?, Tarragona – Catalunya, February 2004. 14 Jan. 2008 <http://www.ciemen.cat/mercator/pdf/simp-toggenburg.pdf>

Van Parijs, Phillippe. "Europe's Linguistic Challenge." *European Journal of Sociology* 45.1 (2004): 111-152.

Van Someren, Emily: "The EU Language Regime: Lingual and Translational Problems." *M.A. dissertation, Bilthoven, Utrecht University, 2004*. 14 Jan. 2008 <http://www.ethesis.net/eu_language/eu_language_contence.htm>

Yves, Peter. "Managing or Celebrating Linguistic Diversity in the EU?" *Institute for European Studies*. 2004. 14 Jan. 2008 <http://www.iee.umontreal.ca/pubicationsfr_fichiers/COLLOQUE-2004/IvesIESfinal.pdf>

Metamorphosis of Civic Identity within Campanella's Walls and on More's Island

Emine Yeşim Büyükaslan

Abstract

Utopian land is a homogenous place where we find, as H.G Wells explains, "a multitude of people, healthy, happy, beautifully dressed, but without any personal distinction whatever" (9). *Utopia* and *The City of Sun* perfectly describe the communal, insular, self-sufficient and changeless identity and culture. The structure and architecture of their utopian societies, cities are identical, and every detail is arranged for the new unified identity within the walls of Campanella's city or on More's island. This paper will show why identity and culture cannot flourish freely, as it should be, in utopia. The identical characters of utopians reflect the metamorphosis thoroughly because being identical is against human nature. Therefore, metamorphosis of each individual affects the blooming of identity and culture.

This article discusses the impact of place and political regime on identity as imagined in two major utopian works: *Utopia* by Thomas More and *The City of Sun* by Tomasso Campanella. Place, identity and political regime are closely related to each other in utopia. These authors conceptualize utopia as a place where humanity experiences metamorphosis of identity as a result of the state's policies. More and Campanella conceive of the state and its location, its place as organized to form a unified identity.

The utopian regime has different policies on social life, economics, education, politics, and family life. The individual's identity emerges and develops as part of a trained social awareness. Utopian policy forms a new identity defined by will of the state rather than that of the individual. Utopian states are based on justice and equality. Therefore, the location of a utopian land must facilitate these ideals. Place has an important role on identity and culture because humanity arranges itself according to geographical features. In this paper I will explain the metamorphosis of human identity as a product of the metamorphosis of the place they inhabit.

Utopia emerges from the infinite desires of humankind. As Busky states, "Desires are infinite in nature" (21), and this is the most distinctive feature of humankind. Human desires have no limits and boundaries. Therefore, utopia emerges from unlimited human desires, and hopes. The ultimate human desire is to reach happiness, and people always long for a better society, especially when the social system does not function well nor satisfy their demands. Every society has failures and hopes, every society unconsciously desires utopia. It is a universal ideal of humanity.

Thomas More introduced the term "utopia" as a literary genre in the seventeenth century with his famous book *Utopia,* which means "no place" in Greek. "Utopia" is nowhere, an imaginary place. According to Kumar, the desire for utopia is the desire "to live in a world that cannot be but where one fervently wishes to be" (1). Utopian land is very much like heaven, where a fantastic landscape is described and praised. Place is the first concern of all utopian authors because the realm's location also projects the identity of that community. Utopian authors locate their "imagined communities" far away from corrupted civilizations in order to protect the ideal state from the harmful effects of other states and nations (Wegner xvi). Wegner points out that, "The utopia's imaginary community is thus not only a way of imagining subjectivity, but also a way of imagining space" (xvii). Subjectivity of thoughts combines with the space, and a new ideal community emerges. The location of the "imagined community" is an indicator of its residents' life style and standards.

Utopia and *The City of Sun* are products of the Renaissance Era, an age of geographical expansion and discovery of new lands. Rees explains, "Classical texts represent origins; Renaissance texts, products of an expanding world geography, create new locations. Together they systematise the genre, so that seventeenth-century and eighteenth-century writers can apply it imaginatively to construct their own alternative fictions" (8). Geographical expansion and discoveries influence utopian literature during the Renaissance. Renaissance authors set their stories in new lands where they hope to found better political and social systems. They imagine that unknown and mysterious places might provide the perfect environment for the best states and political regimes. In *Utopia* and *The City of Sun,* More and Campanella, who are not content with the political and social systems of their own societies, dream far away lands in order to criticize their respective European realities. Geographical expansion and discoveries feed their imaginations, and they locate their ideal states in new lands where more efficient political regimes may flourish.

Utopia and *The City of Sun* are organized perfectly for the communal life style and a unified, homogenous identity. Utopia is located on an island; The City of Sun is located within walls. Gray enlightens us that *The City of Sun* has "too lengthy description of the city itself, and of its walls" (70). More's *Utopia* also contains detailed description of the island and its geography. Both lands are isolated, self sufficient, and closed to any changes and improvements that might risk the political regime and its future. The state must be stable and there must be coherence between the regime and the identity of the individual utopian. Therefore, More and Campanella dream and depict wonderful lands and homogenous identities and societies. Their utopian lands are at the service of utopian regimes based on communal life style.

Utopia and the City of Sun, newly discovered countries, have cities built with identical structures and architecture. The identical urban geographies contribute to the formation of an identical social structure and civilian identity. Gray states that "Life in Utopia is subject to strict discipline: everything is made to a pattern. The cities are numbered; they are at appropriate distances from each other; the number of families in each city and the size of families are regulated" (65). The city itself plays an integral part of the utopian mission. Bonnie Menes Kahn claims that cities have "a purpose, a sense of mission, an idea" (qtd. in Martin-Jordache 1). The mission and purpose of utopian cities are to integrate utopian policy with humanity so as to assure obedient citizens. Utopian cities are established by the state rather than citizens. Citizens do not have any role to build their own places and cities, because everything belongs to the state as George Orwell writes, "Nothing was your own except the few centimeters inside your skull" (27). Citizens belong to the state not only physically but also emotionally and mentally.

We read that on More's island "There be in the island fifty-four large and fair cities, or shire towns, agreeing altogether in one tongue, in like manners, institutions, and laws" (60). He is describing a homogenous island that metamorphoses each individual through its manners, laws and institutions. Human beings have some common features as well as differences and diversities. However, utopian land does not accept a heterogeneous society, rich cultures and traditions, because the state can easily lose its control over the masses. Therefore, society is unified and "the whole island is, as it were, one family or household" as More states in *Utopia* (78). This means that society is not composed of families, but rather society itself is a family, sharing the same identity within the "city home".

The utopians have a monotonous city life. The cities do not have their own atmosphere, architecture, culture and history because, as More visualizes, "they be all so like one to another, as far forth as the nature of the place permitteth" (63). A unified geography and nature accelerates the homogenization of human lifestyle, emotions and reactions. As we all know, one of the basic human needs is security. Without security and peace, human psychology becomes damaged and unhealthy. Fear and insecurity would cause trauma in society. A happy society is a secure society. More's cities in *Utopia* are identical and utopians do not feel insecure. There are no differences between the cities and there is nothing new to explore or cause excitement. More shows us a place where every individual is metamorphosed by the island as communal identity is formed.

The combination and unification of human identity with the island is done by King Utopus. As he conquers the island, he metamorphoses each individual into perfect human beings who have civil gentleness and good fashions (More 60). After structuring homogenous cities, he metamorphoses human thoughts, behaviors and emotions. First, equality must be set among people and their basic needs must be satisfied. Poverty must be abolished from the society, because inequality would also make people feel insecure. For that reason, Utopia has a communist regime and this policy abolishes inequality as it sets a uniform island. The utopians have "all manner of fruits, herbs, and flowers, so pleasant, so well furnished" (More 64). Therefore, people do not commit crimes because they have everything they need and there is no reason for antisocial behavior.

Metamorphosis of each individual continues with clothing each one after the same fashion. Men and women, married and unmarried wear different clothes. More's clothing policy classifies each individual according to gender and social roles. However, the seasonal change is ignored and clothes fit for winter and summer. More tries to abandon diversity even in clothing in order to promote a simple and unified identity for each member of the collective. Communal life style is imposed and there is no room for creativity and imagination. This approach aims to abolish human greed, understood by More as the cause of inequality and unhappiness in a society.

The City of Sun also depicts the same kind of metamorphosis. The city is located within walls. The location of the city is secure for the residents. The people of *The City of Sun* come from India. They escape from a tyrant and settle within the walls to lead a philosophical life. This indicates their unhappiness with past and an ambition to be happy in the utopian present. Aspiration to a philosophical life is a sign of metamorphosis. A philosophical life inspires change and improvement based on the cultivation of conscious. The city is ruled by three virtues: Power, Wisdom and Love. Power relates to war and peace. Wisdom encompasses science, schools and all disciplines. Love is contemplated for its role in procreation. The philosophical life within walls is the first condition necessary for the metamorphosis of Campanella's utopians.

Campanella depicts such a homogenous society that, "Arts and honors and pleasures are common, and are held in such a manner that no one can appropriate anything to himself" (1). *In The City of Sun*, there are no individual pleasures, manners and emotions that distinguish one citizen from another. By and large we understand that diversity is in human nature. However, no one can be different and unique within the restrictive confines of utopian land. The policy aims to create identical human beings who would accept "dwellings in common and dormitories, and couches and other necessaries" (Campanella 1). They own nothing; even their pleasure and emotions belong to the utopian state.

Campanella's walls not only protect the residents but also educate them. The state decorates walls with pictures and formulae of each discipline. Utopian pedagogy emphasizes conditioning and indoctrination. These methodologies metamorphose children from their infancy. Children belong to the state, and the state is responsible for their education because, as Campanella writes, "individuals for the most part bring forth children wrongly and educate them wrongly" (1). The state imposes its ideology on the children so that "each child would learn the utopian principles from infancy and would grow up into a new kind of person dedicated to the new way of life" (Kesten 113). Nobody is allowed to form a unique, personal identity. Education is an important policy used by the utopian regime to construct collectivized citizens.

Campanella and More design societies wherein human beings metamorphose into perfect utopians. These are monolithic societies. As they believe utopia the most perfect society that humanity has never had, they do not allow any alternatives. Therefore, utopia "denies or grasps every alternative, and where meaning is not kept open and the object of debate. Such totalitarian and repressive utopias offer closure and no possibility for future change, only a realization of one grand design. This is Utopian thinking without any possibility of alternatives" (Dahlberg, and Moss 190). A closed society restricts people and impedes critical thinking skills. In fact, the utopian state would aim to abolish critical thinking, because it leads to rebellion, and disobedience. Campanella visualizes absolute obedience of men and women to the singular project as he writes, "It is wonderful to see how men and women march together collectively, and always in obedience to the voice of the king" (1).

More and Campanella support eugenics because, as Booker observes, "marriage is considered a service to the state rather than an expression of individual love" (48). Marriage is arranged between healthy and intelligent males and females for healthy offspring. Campanella claims that "Love is foremost in attending to the charge of the race. He sees that men and women are so joined together, that they bring forth the best offspring" (1). Therefore, romantic love is forbidden, whereas the love for state is appreciated and encouraged. The state cannot tolerate emotional bounds between individuals. Campanella admits that "when we have taken away self-love, there remains only love for the State" (1). The responsibility is towards the state, and its laws, not to the family or other people. Any emotional unity among male and female would lead to the demise of utopian state. Therefore, there is no love, only procreation.

We see *Utopia* and *The City of Sun* as realms of restriction and totalitarianism. All the practices of the utopian state regarding education, marriage, and social life prohibit individuals from forming their own identities. The first restriction is the place they inhabit, because the organization of each city serves the purpose of the utopian policy. Isolated kingdoms, equally divided and structured cities complete the purpose and the willingness of the utopian rulers whose policies eliminate differences and unify a single identity. Thus, identity cannot flourish freely because utopian policy transforms everyone into mechanical and unemotional human beings. People are the products of the same manufacture, and they are equal without distinguishable features. They exhibit nothing original and unique because new ideas and notions potentially damage utopia's peaceful atmosphere.

In conclusion, utopian land is, where we find, as H.G Wells explains, "a multitude of people, healthy, happy, beautifully dressed, but without any personal distinction whatever" (9). They are all happy with who they are, and what they have. In fact, these identical characters reveal the metamorphosis thoroughly because such absolute equality is against human nature. Utopia derives from change and metamorphosis. Metamorphosis is the engine of utopia; according to William Fulbright, utopia does not tolerate human weakness and is unwilling to accept man as he is (79). Humanity's weakness is replaced with self-sufficient and obedient citizens who serve the state in the best way. Not only do they serve and obey but also as Harris put it, "hears of the voice of 'the State' within him" (qtd. in Gleason 17, 18). As the individual hears the voice of the state within himself, he loses the ability to hear himself.

Works Cited

Booker, M. Keith. *The Dsytopian Impluse in Modern Literature: Fiction as Social Criticism*. Westport, CT: Greenwood Press, 1994.

Busky, Donald F. *Communism in History and Theory: From Utopian Socialism to the Fall of the Soviet Union*. Westport, CT: Praeger, 2002.

Campanella, Tommaso. *The City of Sun*. Project Gutenberg. 1. Sep. 2001. <http://www.gutenberg.org/dirs/etext01/tcots10.txt>

Dahlberg, Gunilla, and Peter Moss. *Ethics and Politics in Early Childhood Education*. London: RoutledgeFalmer, 2005.

Fulbright, William J. *Arrogance of Power*. New York: Vintage Books, 1966.

Gray, Alexander. *The Socialist Tradition, Moses to Lenin*. London: Longmans, Green and Co., 1946.

Gleason, Abbott. *Totalitarianism: The Inner History of Cold War*. New York: Oxford
University Press, 1995.

Kesten, Seymour R. *Utopian Episodes: Daily Life in Experimental Colonies Dedicated to Changing the World*. 1st ed. Syracuse, NY: Syracuse University Press, 1993.

Kumar, Krishan. *Utopianism*. Great Britain: University of Minnesota Press: 1991

Martin-Jordache, Corina. "*Modernity, Urban Semiology and the Beckettian Cityscape*." Journal
of European Studies 32.4 (2002): 351+.

More, Thomas. *Utopia*. Ed. Tom Griffith. Trans. Ralph Robinson. Wordsworth: Great Britain, 1997.

Orwell, George. *Nineteen Eighty-Four*. Penguin Books: England, 2000.

Rees, Christine. *Utopian Imagination and Eighteenth-Century Fiction*. New York: Longman, 1996.

Wegner, Phillip E. *Imaginary Communities: Utopia, the Nation, and the Spatial Histories of Modernity*. Berkeley, CA: University of California Press, 2002.

Wells, H. G. *A Modern Utopia*. Lincoln, NE: University of Nebraska Press, 1967.

HEMINGWAY'S HISPANIC VISION IN *FOR WHOM THE BELL TOLLS*

JEFFREY HERLIHY

Abstract

This article examines Robert Jordan's Anglo-American identity within the Iberian context of Ernest Hemingway's Spanish Civil War novel *For Whom the Bell Tolls* (1940). The expatriate protagonist's social, linguistic and national characteristics function as separators while he is in Spain, and the foreign point of reference allows the author to make a range of inferences about social acculturation and national identity. This article interprets how Robert Jordan reacts to the new society during his immersion within the Spanish culture.

The expatriation of the characters in Ernest Hemingway's fiction is a clear binding element in all of his novels, and it is a theme with social implications that critics have overlooked.[1] In *For Whom the Bell Tolls* (1940) Robert Jordan is fighting on behalf of the Republic in the Spanish Civil War, and the American uses the situation in Spain to make forays into foreign behavior such as language, alcohol, and marital rites in order to be a different person abroad than he was at home. The purpose of these cultural ceremonies is to assimilate with the society of the foreign place, and for the actor to achieve a new sense of self. While the trials in non-native behavior seem profound, in the end they are fleeting, for the new social group ultimately rejects the American. I will use an analysis of the protagonist's failed cultural rites of passage in Spain to elucidate how

---

[1] While Robert Martin's "The Expatriate Predicament in *The Sun Also Rises*," "Hemingway's Expatriates" by Nadine Gordimer, and Angel Capellan's *Hemingway and the Hispanic World* discuss the aesthetic results of Hemingway's foreign settings, the studies do not address the protagonists' use of foreign behavior as a social device.

Robert Jordan's foreign situation affects his actions, self-image, and perspective of the world around him.

Hemingway explored Europe between 1918-29 first as an ambulance driver in the Great War, then as a newspaper correspondent, and finally as a novelist. The raw material from his European experience allowed him to create in his fiction a world apart from his Midwestern background; this extra world in his writing provides a source of new values which became part of each character's psyche and consciousness. More specifically, the perspective of an American in Europe is in constant comparison with the cultural and social codes of Europeans, such as military men, prostitutes, bartenders, and *toreros*. The condition of foreignness, actor out of context, is the creative vehicle which Hemingway employs in every novel and many pieces of short fiction to produce dramatic tension.

As a result, Robert Jordan's experience in Spain is part of a series of recurring over-arching themes in Hemingway's novels. Hemingway's archetypal expatriate protagonist has common characteristics which we see in each novel: the main character is an expatriate, bilingual,[2] adult male who has lived several years abroad when the drama begins. During the years living abroad prior to the action of the novel, the actor adopted cultural elements from the foreign societies in language, sports, alcohol, and religious ceremony among other things, and developed them into his lifestyle. The purpose of these actions in the foreign culture is to pass into the society of the new place, and achieve a new sense of identity. The protagonist actively avoids compatriots, and, in Hemingway's first three novels set in Europe, the protagonist believes he has forged a transcendent kinship with the society of the new place. In *The Sun Also Rises* (1926) Jake Barnes calls this kinship "a special secret" (131); in *A Farewell to Arms* (1929) Frederic Henry names it "a funny sort of life" (26); and for Robert Jordan in *For Whom the Bell Tolls* (1940) it is "a strange sort of life" (165), "completely integrated" (471). As we can see, each protagonist defines himself in terms of his relationship with a foreign people, and the foreign social context functions as a tool which Hemingway uses to demonstrate each man's sense of being.

Several anthropologists, psychologists and cultural theorists have examined the results of prolonged expatriation on a person's sense of self; these studies demonstrate that people that are out of their native social and

[2] Santiago in *The Old Man and the Sea* is monolingual; however, the protagonist that emigrated from the Canary Islands to Cuba speaks two dialects of the Spanish language.

cultural contexts for an extended period of time act in predictable ways.[3] The social isolation which the exile endures causes disproportionate pressures on the individual, and these strains trigger negativism and depression. Consequently, expatriate unhappiness is universal in Hemingway's texts, as Jake Barnes declares about Paris in an unpublished manuscript of *The Sun Also Rises*: "there is nothing romantic about the Quarter and very little that is beautiful" (JFK MS 194-10). Edward Said remarks that people in exile have "an urgent need to reconstitute their broken lives, usually by choosing to see themselves as a part of a triumphant ideology or a restored people" (177). Said goes on, explaining that the expatriate "is offered a new set of affiliations and develops new loyalties [with another social group]" (183). Exactly how an individual establishes loyalty and confidence in association with another society is a topic developed in Arnold Van Gennep's monograph *The Rites of Passage* (1960). The study finds that for an individual to pass from one social group to another, he or she must fulfill a series of ritualistic trials, which are, in a sense, affirmations of loyalty, before the actor can be a member of the new group. The purpose of these ceremonies is to move from one defined social position to another (1-2). Thus, for an individual to enter in one society from another, there are a series of rituals particular to each culture that the "other" must perform. The purpose of these rituals is to differentiate the person from his or her background, to grant oneself a *tabula rasa,* so to speak. The person "must prove intentions from afar and undergo a transitional period" (28). This period is vastly varied from society to society, and it may include the exchange of gifts, gestures, lodging, linguistic exchange, and informational exchange. The speaking of a foreign tongue is a fundamental method of joining oneself to a new group; "The simple fact of pronouncing a word or a formula," says Gennep, "has the effect of creating [...] bond" (32).

As I remarked, the archetypal Hemingway protagonist formed his relationship with the new society prior to the start of each text, and Robert Jordan is no exception. Hemingway presents the protagonist *in medias res* of the actor's foreign life, and to analyze how the actor came to identify with the other society, we must work backwards, using the scant textual clues that refer to the character's personal history abroad. In *For Whom*

[3] This circumstance results in higher rates of alcoholism, domestic abuse, and admission to mental hospitals for immigrants than nationals (Packard 232). Some valuable peripheral studies of place, exile, and the psychology of dislocation are: *Place and Placelessness* by E. C. Relph; Leonard Lutwack's *The Role of Place in Literature*; *Imagining Paris: Exile, Writing, and American Identity* by J. Gerald Kennedy; and *Space and Place: The Perspective of Experience* by Yi-Fu Tuan.

the Bell Tolls, the protagonist is a man from Montana, who is fighting with a Castilian guerrilla band during the Spanish Civil War. Prior to the text, Jordan spent parts of a decade traveling in Spain and during these years he learned the Spanish language and became knowledgeable on Iberian customs (135). It was the stresses of the prolonged expatriation and its accompanying social isolation which caused Robert Jordan to adopt foreign cultural practices, and the man uses the Spaniards as a surrogate community to alleviate the pressures of dislocation.

A principal method Jordan uses to connect with the Castilian people is through use of their language. Several critics have examined the Spanish-language elements in *For Whom the Bell Tolls* (see Fenimore 73-86; Josephs 221-41; Fleming 265-77). While many scholars comment on the realism, the accuracy of the orthography, and aesthetic results of Hemingway's Spanish linguistic experiment, few academics recognize the bonding motif which Jordan achieves with the Spanish people through his use of their language. Two critics, Edward Fenimore and Mimi Gladstein, touch on the uniting quality that the American's Spanish-language ability has with the people of the Republic. Gladstein writes that use of foreign language displayed "affinity to the people they were depicting, to position [Jordan] with, rather than above, the working class. In this living within, between, and sometimes outside of two cultures" (81). Fenimore relates that the use of Castilian shows "the non-Spanish looking in on the Spanish world" (73). A principal objective of foreign language in the novel is to demonstrate the integration that Jordan achieved through his years immersed in the other society.

Robert Jordan is a foreigner in Spain acting as a military supervisor on behalf of the Republican army, and his authority within the social circle of Spanish peasants relates directly to his bilingualism. Jordan remarks "they trusted you on the language, principally. They trusted you on understanding the language completely and speaking it idiomatically" (135). When the American first arrives with the peasant group, their archaic dialect of the Spanish language confuses him, and Jordan consciously alters his own use of Castilian in order to amalgamate with the new community. Anselmo spoke "in a dialect that Robert Jordan could just follow. It was like reading Quevedo. Anselmo was speaking old Castilian and it went something like this, 'Art thou a brute? Yes. Art thou a beast?'" (11). Up until this point, Jordan has spoken only in modern Spanish dialect devoid of 'thees' and 'thous'—and the American *changes* his own Spanish from the standard dialect adopting the old parlance in order to better associate with Castilians; the initial time Jordan employs this new variant in his Spanish is the first time he directly addresses Maria, saying: "How art thou called?" (23). These linguistic forays not only in Spanish, but also into the archaic dialect function in Jordan's social favor, for, partly through his language skills, he achieves a romantic bond with the peasant girl. The relationship with Maria, which I will discuss later, is an essential building block in Robert Jordan's foreign identity.

At one point, El Sordo, the leader of another guerrilla band, uses pidgin Spanish with Jordan because he is a foreigner. Sordo's first phrases to the American are not sentences but single words accompanied by hand gestures: "'Eat,' he said and looked at Robert Jordan. 'Drink?' he asked, making a motion with his hand decanting his thumb downward" (141). This basic talk embarrasses the American, since he already proved his language skills with the other band of Castilians. "Does he talk that way to every one? Robert Jordan thought. Or is that his idea of how to make foreigners understand?" (146). El Sordo humiliates Jordan and he wishes he would not "speak that pidgin Spanish for foreigners" (147). Later in the text, Sordo speaks with regular fluency to Jordan—using the same speed and vernacular as though he were speaking with another Spaniard—a detail which demonstrates the trust which the American gained with the new group.

The social element of Jordan's linguistic ability cannot be overstated: "He speaks Spanish as we do" (209) remarks Anselmo, which, like Sordo's transition from single-words to standard spoken Spanish, eventually allows Jordan to believe he had "an absolute brotherhood with the others" (235). As the narrator observes toward the end of the novel, Jordan felt "completely integrated now" (471) as he lay with his heart on the Spanish earth after blowing the bridge. Jordan believes his achievement in Spain transcends the destruction of a river overpass on behalf of the Republic; the American created a social identity for himself through rites of passage, and this new sense of self eventually supersedes his American identity, for in the end of the drama, Jordan is willing to die for the new society.

Another important concept in this novel is Robert Jordan's objective to "live a life in two days" (169) while he fights in the mountains with the rebel band. An essential ingredient in this "Spanish life" is his marital engagement with Maria, a member of the foreign society. In a similar vein to the author's other novels,[4] in *For Whom the Bell Tolls*, Ernest Hemingway uses an American man's intention to wed a foreign woman as

[4] In *The Sun Also Rises*, *A Farewell to Arms*, and *Across the River and into the Trees*, each of Hemingway's male protagonists has the desire to wed a foreign woman, and in *The Old Man and the Sea*, Santiago, the Spaniard, was married to a Cuban woman.

a tool to develop character. In this Spanish Civil War novel, Robert Jordan's intention to marry Maria serves to facilitate the man's acculturation process in the new country. While Robert Jordan's knowledge of the Spanish language and cultural rites with food and alcohol integrate him with the band of rebels in the mountains, it is the man's intention to realize betrothal rites with Maria which concretizes the American's transformation. The ritual allows the foreign group of rebels to accept Robert Jordan into their society through a ceremonial process, and while the matrimonial rites do not occur, the notion of a social bond between the American and the Spanish woman is an important element of Jordan's character.

The marital commitment between Jordan and Maria transpires very early in the text when Maria says to Jordan after their first sexual encounter: "I can be thy woman" (73). Throughout the novel Hemingway uses the technique of literally translating spoken dialogue from Spanish into English, and in this way, Hemingway uses the word "woman"— to identify Maria as Jordan's wife.[5] Just after this scene, Jordan and Maria express themselves publicly to the group as an engaged couple, and the Spaniards' positive reception of the engagement is a principal element of Jordan's acculturation. In Chapter 9 Pilar announces to the others: "That comrade is her *novio*[6]" (92). And Fernando, who initially feels uncomfortable at Jordan and Maria's public displays of affection, remarks that "[s]ince they are engaged, I encounter it to be perfectly normal" (92). Pilar responds "I am pleased" (92). Fernando then toasts Pilar on this acknowledgment. The collective recognition which the Spaniards display toward Jordan's marital intentions with Maria is a vital part of the American's sense of self in the foreign country.

Following their nuptial agreement, Robert Jordan's self-image distinctly changes; for example, after the dialogue about their engagement, Maria calls Jordan "*Inglés*"—and he corrects her, saying "My name is Roberto." The Castilianization of his own name to "Roberto" affirms that the American believes his identity has transformed, and he therefore conceptualizes himself with another, Spanish, title. "Nay," responds Maria. "I call thee *Inglés.*" "Still it is Roberto," he responds. Maria disagrees again, saying "[n]o." Jordan replies "What I do now I do alone" (172). It is very clear Jordan's "Spanish self" is important, and he wants Maria to acknowledge that the couple, due to their engagement, are no longer foreigners to one another. Jordan's regression into the first person

---

[5] In Spanish, *mujer* means woman as well as *wife.*

[6] *Novio* translates to English as *boyfriend, fiancé,* or *groom*, depending on the context.

pronoun in this specific interchange illustrates Jordan's feeling of separation, and foreshadows his epiphany later in the novel that the Spaniards in fact reject the American. Hemingway uses Robert Jordan's engagement with Maria as a method to illustrate the American's integration with the Spanish society, and the profound reaches of matrimony evidence Jordan's dedication to being a member of the new social group. The affirmative reactions from the other members of the guerrilla band strengthen Jordan's feeling of transition, and the social rite of betrothal allows the American to solidify his new identity.

From this effort to assimilate in language and culture, Jordan believed he was "in as far as any foreigner could be. He never felt like a foreigner in Spanish and they did not really treat him like a foreigner" (135). Like Jake Barnes and Frederic Henry, who go through similar experiences in Hemingway prior novels, Jordan's rites of passage in the new culture allow him to feel like a national of a foreign country, and together with the tale of guerrilla fighting in the Spanish war, and the consequences of fascism spreading around Europe, *For Whom the Bell Tolls* is also a story about a man's feeling of acceptance within an adopted community. One of the chief aims in *For Whom the Bell Tolls* is to relate the description of an American making an affiliation with a foreign people, and the rejuvenation the man experiences from such a process.

What is the final outcome of this American expatriate acculturation process? Despite having feelings of renewal, the ultimate result of forging a place in another society in the drama is irony and sadness, since at the end of Hemingway's war novel, the other society does not completely accept the foreign protagonist. While the Spanish land has given Jordan a new sense of self and an ostensible place within a new social circle, in the end his achievement is unclear, since his death is on behalf of a cause which does not have the transcendence he tries to apply to it.

On several occasions Jordan reflects on the failure of his Spanish situation; the American understands it is a mistake to idealize the relationship with the others. "Don't go romanticizing them" (204) he says to himself, and due to his frustration with the group, he regresses from time to time into the English language. "'Why do you speak in English?' Pablo asked. 'I don't know,' Robert Jordan said. 'When I get tired I speak English. Or when I get disgusted'" (180). Jordan is not disillusioned from the preparations to destroy the bridge, but with the effort he must make to feel at home with the Castilians. In fact, many of the Spaniards are unsympathetic to Jordan throughout the novel: "This foreigner comes here to do a thing for the good of the foreigners," Pablo remarks (54). And Fernando observes "It is, in a way, presumptuous for a foreigner to teach

Spanish" (209). Pilar attempts to defend the American, saying "'Can't you hear he speaks Spanish?' 'Yes,' says Fernando, 'But with an accent'" (210). Moreover, the members of Castilian guerilla band repeatedly refer to Jordan as "*Inglés*" or English as a playful method of establishing social positions. Even after Jordan explains he is American, not English, the Spaniards continue using the nickname—emphasizing the solidarity of the Spanish community, and the foreignness of Jordan. His relationship with them, while warm at times, is always that of an outsider.

Despite the "brotherhood" Jordan believes he shares with the fighting band, in reality the American spends "not quite three days and three nights" in the mountains, and the attachment he shares with the other group is fleeting (466). The truth of the situation is that Jordan dies alone, far from his family, in a country where his best connections are with a people the American has known under one week. While the foreign land has given Jordan a new sense of self and an ostensible place within another social circle, in the end his social success is ambiguous—for the Spaniards do not accept him.

This extraneous social element of the storyline in a sense enriches the plot, giving a human dimension to the protagonist—but Jordan's fraternizing with the Spanish changes his perspective in an unreal way; he is never one of them, and his stated purpose in Spain is: "I am going back and earn my living teaching Spanish as before, and I am going to write a true book" (163). Jordan's motivation from the beginning is to return to Montana to teach and write, making the experience in Spain at bottom an aesthetic and economic one, with clear ties to aspirations back in his native land. It is Pilar, more than the other Spaniards, who is keen to Jordan's ultimate purpose with them in the mountains, and the woman makes it clear she knows Jordan's real intentions. "Shut up about what we are going to do afterwards," she says. "You take your piece [Maria] with you and leave us others alone here" (150).

Prior to the operation at the bridge, Jordan becomes dejected about his position as an outsider: "Muck every one of them to death and to hell. Muck the whole treachery-ridden country. Muck their egotism and their selfishness and their selfishness and their egotism and their conceit and their treachery. Muck them to hell and always. Muck them before we die for them" (369). It is only once it becomes evident that Jordan will not survive the ordeal at the bridge that he turns to the unreal idea of "integration" with Spain. At the closing moments of the text, the American applies this social angle to the dire situation in order to alleviate the intense pressure of the moments preceding his death.

"What a business," Jordan remarks. "You go along your whole life as though they mean something and they always end up not meaning anything. There never was any of what this is" (167). The protagonist comprehends the meaninglessness of the foreign enterprise, and also the failure of his interaction with the Spanish when Jordan says "they did not really treat him like a foreigner most of the time; only when they turned on you. Of course they turned on you. They turned on you often" (135). And later Jordan consents that there are "no worse people in the world. No kinder and no crueler. And who understands them? Not me" (355). Jordan's amalgamation with this society he does not understand is romantic and artificial; the American is wise to the reality that a "Spaniard was only really loyal to his village in the end" (135). The final irony of Jordan's status in Spain is foreshadowed very early in the text, when Jordan remarks "[t]hat I am a foreigner is not my fault. I would rather have been born here" (15).

Edward Said points out that "Exile is strangely compelling to think about but terrible to experience. It is the unhealable rift forced between a human being and a native place, between the self and its true home: its essential sadness can never be surmounted" (173). Once an actor is abroad beyond a stint of tourism, Said goes on, "he must deal with the crippling sorrow of estrangement" (173). To compensate for these feelings of loss which are typical during expatriation, Hemingway's protagonist in Spain fashioned a new persona within the foreign society, but the reaches of Jordan's endeavor are superficial. As Robert Jarrett deftly explains, "identity cannot be maintained solely by one's own power; the choice of identity is limited by the influences of personality, geography, and culture," he goes on, "crossing the border allows people to strip away layers of historical identity, familial, and the person, only to discover that across the border authentic identity can only be lost, not found" (120). In the case of Robert Jordan, to be one of them in Spain the actor had to cease being an Anglo in culture, in language, and in being—a task which was not possible. Due to the inconsistency in the human social group, something beyond the control of the character, the foreign society does not unconditionally accept him. Robert Jordan could not, in the end, choose his identity.

On the last page of the Hemingway's Spanish Civil War novel, Robert Jordan lay on the pine needles of the forest floor waiting to die. The omniscient narrator observes that the American "was completely integrated now and he took a long look at everything" (471). This phrase stands alone as a climactic and paradoxical moment in the Robert Jordan's "Spanish life"; the American man came to Spain to compile aesthetic observations for a book and for the background experience for the classes he would teach after his return to Montana. The text ends with the American man about to lose his life on behalf of a loose band of foreigners who, despite the outsider's attempts in Spanish cultural rites, do not accept Jordan in their society. The group ultimately leaves the injured American behind to die so that they may escape.

In *For Whom the Bell Tolls* Ernest Hemingway uses a foreign setting as a literary resource to demonstrate Robert Jordan's altruism, compassion, and desire for personal cultivation. The Hispanic context of the novel is one of the principal vehicles which the author uses to build and develop character, and through scrutinizing Robert Jordan's interaction with the people of Spain, we perceive a transformation in the man's sense of himself. While the American believes that his bilingualism and intention to wed a Spaniard will make him a member of the new society, in the end the new community rejects Robert Jordan, and this social failure is a principal stress in the American's situation abroad. Hemingway's use of a foreign setting and the protagonist's interactions with another society are essential components to the drama itself, and our comprehension of Robert Jordan's character depends to a great extent on the man's attempt and ultimate failure to integrate with a new social group.

In Hemingway's own words, "Americans are always in America, whether they call it Paris or Paname" (Selected Letters 218).

Works Cited

Capellan, Angel. *Hemingway and the Hispanic World.* Ann Arbor: UMI Research Press, 1985.

Fenimore, Edward. "English and Spanish in *For Whom the Bell Tolls.*" *ELH* (March 1943): 73-86.

Fleming, Bruce. "Writing in Pidgin: Language in For Whom the Bell Tolls." *Dutch Quarterly Review of Anglo Letters* 15:4 (1985): 265-77.

Gennep, Arnold Van. *Rites of Passage.* London: Routledge and Kegan Paul, 1960.

Gladstein, Mimi Reisel. "Bilingual Wordplay: Variations on a Theme by Hemingway and Steinbeck." *The Hemingway Review* (Fall 2006): 81-98.

Gordimer, Nadine. "Hemingway's Expatriates." *Transition* (1999): 86-99.

Hemingway, Ernest. *A Farewell to Arms.* New York: Charles Scribners's Sons, 1929.

—. *Across the River and into the Trees.* New York: Charles Scribners's Sons, 1950.

—. *For Whom the Bell Tolls*. New York: Scribners, 1940.
—. *Selected Letters*. New York: Charles Scribners's Sons, 1982.
—. *The Old Man and the Sea*. New York: Charles Scribners's Sons, 1952.
—. *The Sun Also Rises*. New York: Charles Scribners's Sons, 1926.
—. "The Sun Also Rises Manuscript." *Hemingway Archive*. John F. Kennedy Library, Boston. Item 194 10.
Jarrett, Robert. *Cormac McCarthy*. New York: Twayne, 1997.
Josephs, Allen. "Hemingway's Spanish Sensibility." Ed. Scott Donaldson. *The Cambridge Companion to Ernest Hemingway*. Cambridge: Cambridge University Press, 1990. 221-41.
Kennedy, J. Gerald. *Imagining Paris: Exile, Writing, and American Identity*. New Haven: Yale University Press, 1993.
Lutwack, Leonard. *The Role of Place in Literature*. Syracuse: Syracuse University Press, 1984.
Martin, Robert A. "The Expatriate Predicament in *The Sun Also Rises*." Ed. J. Gerald Kennedy and Jackson Bryer. *French Connections: Hemingway and Fitzgerald Abroad*. London: Macmillan, 1998. 61-73.
Packard, Vance. *A Nation of Strangers*. New York: Mckay, 1972.
Relph, Edward. *Place and Placelessness*. London: Plon, 1976.
Said, Edward. *Reflections on Exile*. Cambridge: Harvard University Press, 2000.
Tuan, Yi-Fu. *Space and Place: The Perspective of Experience*. Minneapolis: University of Minnesota Press, 1977.

WE'LL ALWAYS HAVE PARIS: METAMORPHOSED PLACES IN VICTOR HUGO'S *NOTRE DAME DE PARIS* AND WOLFGANG SEBALD'S *AUSTERLITZ*

ENRIQUE BANÚS

Abstract

What might a profoundly Romantic novel like Victor Hugo's *Notre-Dame de Paris* (1831) and a possibly post–modern novel like Wolfgang Sebald's *Austerlitz* (2001) have in common? There is a considerable distance between both authors and their respective literary works. But in both novels, space plays a fundamental role – and a city becomes the symbol for the metamorphosis of places.

Victor Hugo's Sorrows

Notre-Dame de Paris is a made-to-order novel, written in two years' time. At the end of 1828, the editor Gosselin entrusted Hugo with writing a historical novel *à la* Walter Scott. *Notre-Dame* is, thus, a historical novel: inspired in the time of Louis XI and set in the fourteenth century – the end of the Middle Ages and the beginning of the Renaissance. It was not until 1830 that Hugo began writing his novel. The editor put pressure on him, but due to the revolutions in Paris in 1830, Hugo obtained an extension, so that he published it in 1831 and added some "*inédits*"[1] chapters in 1832.

[1] "Les chapitres ajoutés à cette édition ne sont pas *nouveaux*. Ils ont été écrits en même temps que le reste de l'ouvrage [...] S'ils n'ont pas été publiés dans les précédentes éditions du livre, c'est par une raison bien simple. À l'époque où *Notre-Dame de Paris* s'imprimait pour la première fois, les dossiers qui contenait ces trois chapitres s'égara. [...] Il eût trouvé court de faire un nouveau roman".

In 1829, Victor Hugo wrote a letter to his editor Gosselin where he affirmed *Notre-Dame* to be "a painting of Paris" – the space is, accordingly, essential. But Hugo goes on saying that "[the book] is a painting of fifteenth century Paris and also a painting of the fifteenth century regarding Paris": is time essential? And if so, what time? One critic states that "What *Notre-Dame* really pretends is to be a mirror of the social and politic reality of the first half of nineteenth century France" (Torrego 25). In any case, places, buildings, play a fundamental role – and many times the author mentions their transformation from the fifteenth century to Hugo's time.

An example: the Palace of Justice, built in the time of Saint Louis, represented one of the most emblematic monuments of fifteenth century Paris. Inside, its Gothic vault filled with moulds stands out, just like its immense pillars, sustaining the roofing; one could also admire the enormous square room with its famous marble, round which highly relevant historical events took place. The superior side of the wall was surrounded by sculptures of the kings of France. The stained-glass windows next to the annex chapel gave the Palace a special charm (Hugo, 9-23, Book 1, chapter 1).

But the building underwent various transformations: first of all, the fire of 1618 which destroyed it (Hugo, 12). Even worse are the unprofessional restorations which were done in order to save the building: "Grâce a cette catastrophe, grâce surtout aux diverses restaurations successives qui ont achevé ce qu'elle avait épargné, il reste bien peu de chose de cette première demeure des rois de France, de ce palais aîné du Louvre [...] Presque tout a disparu" (13).

This space suffers important metamorphoses in the novel also, and these metamorphoses are to some extent a tough critique of the contemporaneous society. Neither the aristocracy, nor the Flemish ambassadors, or the bishops, not even the Paris cardinal escape the critics. The people go to the Palais in order to see the representation of the medieval mystery of *Le bon jugement de Mme la Vierge*. The act which turns the Palace of Justice into a grotesque place occurs when the Parisians, in complicity with the Flemish ambassadors, name Quasimodo Pope. The chapel of Saint Louis ends up filled with blasphemous mob[2].

Though Hugo thinks that "son des chapitres d'art et d'histoire qui n'entamaient en rien le fond du drame du roman," (Hugo 5-6) they are essential for this paper.

[2] "La table de marbre, l'estrade de brocard avaient eu leur moment; c'était le tour de la chapelle de Louis XI. Le champ était désormais libre à toute folie. Il n'avait pas plus qui Flamands et de la canaille" (48).

Another symbolic transformation of the Palais de Justice takes place when innocent Esmeralda is condemned to death by hanging, accused of murder and witchcraft. The Palais de Justice looses its moral legitimacy. There are no more differences between it and the suburbs of Paris, where the leprous rob, the crippled assassinate, women sell their bodies and old women cook kidnapped children. The gallows reign there, too: we are in La Court des Miracles (84). The place is transformed not so much by the passing of time but by human action.

Hugo undoubtedly satirizes those who feel pleased with the fact that from the impressive Place de la Grève, splendorous in the fifteenth century, a sinister element has disappeared: the gallows, which was where bandits were hanged.[3] The old splendour is gone: "Il ne reste aujourd'hui qu'un bien imperceptible vestige de la place de la Grève telle qu'elle existait alors" (59), and so are the gallows.[4] One could argue that the nostalgia should not be very strong. But what is the equivalent of the old gallows? "Une misérable guillotine, furtive, inquiète, honteuse, qui semble toujours craindre" (61). There is no room for a positive vision: recent history has not improved the sinister elements belonging to the past, it only annihilated its beauty.

Not only specific buildings capture the author's attention: the second chapter of the fifth book is entitled "Paris à vol d'oiseau." In it, the entire city and the changes it underwent are described. In the fifteenth century, with less than half a million inhabitants, "Paris était encore divisé en trois villes tout à fait distinctes et séparées, ayant chacune leur physionomie, leur spécialité, leurs mœurs, leurs coutumes, leurs privilèges, leur histoire: la Cité, l'Université, la Ville" (117).

It was "une belle ville; c'était une ville homogène, un produit architectural et historique du moyen âge, une chronique de pierre" (Hugo 132), filled with "son originalité, sa raison, son génie, sa beauté" (120): harmony and originality are, thus, its characteristics – lost, from Hugo's perspective, because of the actions generated by the Renaissance, which destroyed the Gothic city and substituted it with a city without personality. Posterior changes continued on this line:

> La renaissance ne fut pas impartiale; elle ne se contenta pas d'édifier, elle voulut jeter bas. […] Depuis, la grande ville a été se déformant de jour en jour. Le Paris gothique sous lesquels s'effaçait le Paris roman s'est effacé

---

[3] In the novel, it is in this square that Esmeralda is hanged.

[4] "C'est une idée consolante, disons-le passant, de songer que la peine de morte, qui, il y a trois cents ans, encombrait encore de ses roues de fer [...] n'ait plus dans notre immense Paris qu'un coin déshonoré de la Grève" (61).

en tour. Mais peut-on dire quel Paris l'a remplacé? [...] Le Paris actuel n'a donc aucune physionomie générale. C'est une collection d'échantillons de plusieurs siècles, et le plus beaux ont disparu. (133-34)

The new houses destroy the old monuments. Because of this eagerness for mediocre renewal, history looses visibility; this is what Hugo deplores: "Du train dont va Paris, il se renouvellera tous les cinquante ans. Aussi la signification historique de son architecture s'efface-t-elle tous les jours" (134).

Undoubtedly, the Cathedral of Notre-Dame of Paris is the most emblematic space of the novel. Around it, Hugo develops with more profusion his nostalgia towards the Middle Ages and his critics towards the changes time operated. It is in the cathedral that all spaces join together. From its towers, dumb witnesses of the sad destiny of the characters, one can watch all the places mentioned in the novel.

Hugo looks up to its facade (he speaks of "parties harmonieuses d'un tout magnifique"[5]) and he describes, like an impressionist painter, the effects of the light[6] and of the darkness[7], the specific moment in time. This cathedral turns into a symbol: "ce que nous disons de l'église cathédrale de Paris, il faut le dire de toutes les églises de la chrétienté au moyen âge" (107), all of them marked by two elements: "variété, éternité" (107).

But the cathedral, too, has deteriorated: time, political and religious revolutions, and the mutilations, dislocations and additions made by the architects defending the so called *good taste* meant its decay[8]. This

---

[5] The author praises it extensively: "Vaste symphonie en pierre, pour aussi dire; œuvre colossale d'un homme et d'un peuple [...] En un mot, puissance et féconde comme la création divine dont elle semble avoir dérobé le double caractère : variété, éternité" (107).

[6] "On était aux premiers tours de mars. [...] Dans ces jours de clarté, de chaleur et de sérénité, il y a une certaine heure surtout où il faut admirer le portail de Notre-Dame. C'est le moment où le soleil déjà incliné vers le couchant, regarde presque en face la cathédrale. Ses rayons, de plus en plus horizontaux, se retirent lentement du pavé du place, et remontent le long de la façade [...] tandis que la grande rose centrale flamboie comme un œil de cyclope enflammé des réverbérations de la forge" (235).

[7] "La cathédrale était déjà obscure et déserte. Les contrenefs étaient pleines de ténèbres, et les lampes des chapelles commençaient à s'étoiler, tant les voûtes devenaient noires. Seulement la grande rose de la façade, dont les milles couleurs étaient trempées d'un rayon de soleil horizontal, reluisait dans l'ombre" (252-53).

[8] "On peut distinguer sur sa ruine trois sortes de lésions [...]: le temps d'abord [...] ensuite, les révolutions politiques et religieuses [...] enfin, les modes, de plus en plus grotesques et sottes, qui depuis *la renaissance,* se sont succédé dans la

architectural decay is only expressing the evolution of the people, since for Hugo "l'architecture a été la grande écriture du genre humain" (177), mankind always passing from theocracy to democracy[9]. In the Middle Ages, this meant abandoning "cette mystérieuse architecture romane, sœur des maçonneries théocratiques de l'Égypte et de l'Inde, emblème inaltérable du catholicisme pur, immuable hiéroglyphe de l'unité papale" (177-78) and moving to the Gothic style: "La cathédrale elle-même, cet édifice autrefois si dogmatique, envahie désormais par la bourgeoisie, par la commune, par la liberté, échappe au prête et tombe au pouvoir de l'artiste. L'artiste la bâtit à sa guise. Adieu le mystère, le mythe, la loi. Voici la fantaisie et le caprice" (178).

The next step presupposes decay: "Mais, dès le seizième siècle, la maladie de l'architecture est visible; elle n'exprime déjà plus essentiellement la société ; elle se fait misérablement art classique ; de gaulois, d'européenne, d'indigène, elle devient grecque et romaine, de vrai et de moderne, pseudo-antique. C'est cette décadence qu'on appelle renaissance" (183).

The Renaissance coincides with the split of religious unity. Protestant reform has its demolishing effect due to printing[10], and this effect, too, can be seen in architecture by Hugo:

> Cependant, quand le soleil du moyen âge est tout à fait couché, quand le génie gothique s'est à jamais éteint à l'horizon de l'art, l'architecture va se ternissant, se décolorant, s'effaçant de plus en plus [...] Elle se dépouille, elle s'effeuille, elle maigrit à vie d'œil. Elle est mesquine, elle est pauvre, elle est nulle. Elle n'exprime plus rien, pas même le souvenir de l'art d'un autre temps (184).

Printing, with its flattening effect, will end up replacing the architecture, which suffers from "minimalization and exhaustion":

> Cependant, que devient l'imprimerie? Toute cette vie qui s'en va de l'architecture vient chez elle [...] Aussi dès le seizième siècle la presse, grandie au niveau de l'architecture décroissante, lutte avec elle et la tue. Au dix-septième, elle est déjà assez souveraine [...] assez assise dans sa victoire pour donner au monde la fête d'un grand siècle littéraire. Au dix-

---

décadence nécessaire de l'architecture. Elles […] ont coupé, taillé, désorganisé, tué l'édifice […] Et puis, elles ont refait; prétention que n'avaient eue du moins ni le temps, ni les révolutions. Elles ont affrontément ajusté, de par *le bon goût*" (109).

[9] "Toute civilisation commence par la théocratie et finit par la démocratie" (177).

[10] "Avant l'imprimerie, la réforme n'eût été qu'un schisme, l'imprimerie la fait la révolution" (Hugo 184).

> huitième siècle [...] elle ressaisit la vieille épée de Luther, en arme Voltaire, et court, tumultueuse, à l'attaque de cette ancienne Europe dont elle a déjà tué l'expression architecturale. Au moment où le dix-huitième siècle s'achève, elle a tout détruit. Au dix-neuvième, elle va reconstruire" (185-86).

In this novel, Hugo grants a crucial role to the *space*, which is more than a simple static or decorative element where the novel's plot develops: space turns into the protagonist of the story. The space in *Notre-Dame* is filled with symbolic elements. And this complex space suffers profound transformations – Hugo uses an omniscient narrator to convey these changes, a narrator who describes the same space in its successive evolutions during the years. The metamorphoses are a signal of profound global changes, contemplated not so much nostalgically, but with resignation: they follow an inexorable law. The result is that Victor Hugo's present time will only be able to rebuild, an activity which is devoid of creativity and originality. Contemplating spaces from the past produces admiration; their transformation, sadness. We have here a profound disappointment, a fatalist vision of history.

Austerlitz' Sorrows – Sebald's Sorrows?

170 years later, Paris and the transformation of some of its spaces – especially one – will play a relevant role in a novel which someday might be considered as representative as *Notre Dame* is of the Renaissance which reinvents the Middle Ages, seen as a mirror in which not only present circumstances can be reflected, but also the passing of time: *Austerlitz*. This is the last novel of W.G. Sebald, German writer and literature teacher at the University of Norwich, deceased in a car accident in December 2001.

The novel, published the same year, recounts a series of encounters that, during the years and with large periods of interruption, take place between the narrator, a character of which the reader scarcely gets any information and the protagonist, Jacques Austerlitz, a former University professor who fills his days travelling through various European capitals, in a hieratic attempt to disinter history – apparently, the history of architecture, although in the end, the reader realizes it was the history of his own past.

Austerlitz is, as the narrator finds out after many conversations with him, one of the many Jewish children who in 1939, before the imminent arrival of the Nazi troops to Prague, is embarked on a ship to Great Britain. He is adopted by a Welsh couple formed of the Calvinist preacher

Emyr Elias and his melancholic wife Gwendolyn. It is not until his adolescence that he finds out these are not his real parents and that the name he was using was not his real name: during his school years, the people in charge of the school see themselves under the obligation, because of legal reasons, to reveal his authentic name, Jacques Austerlitz, although they are not able to provide him with any more data regarding his identity.

Once he finishes high school, and after some years of study in France, he settles in London, city in which he will work as a university professor. The main character's adult life goes by with a conscious and desired self-oblivion. With all his energy focused on absurd research with the purpose of writing a complete history of European architecture, Austerlitz submerges into a deliberate amnesia which allows him to live on the edge of reality. His investigations of the past, whose only purpose is that of befuddling his mind, stop when reaching the nineteenth century, since from this moment onwards, History begins to reveal aspects that make him feel profoundly uneasy.

This situation, of "ignorance through erudition," blows up during the summer of 1992, when a nervous breakdown compels him to spend a long time in bed:

> Yet this self–censorship of my mind, the constant suppression of the memories surfacing in me, Austerlitz continued, demanded ever greater efforts and finally, and unavoidably, led to the almost total paralysis of my linguistic faculties, the destruction of all my notes and sketches, my endless nocturnal peregrinations through London, and the hallucinations which plagued me with increasing frequency up to the point of my nervous breakdown in the summer of 1992 (198).

Only a few months afterwards, once his recovery begins, Austerlitz enters a bookstore and while he skims through some of the books, he listens on the radio to a couple of Jewish women speaking of how in 1939, while they were very young, they had been sent to England in a special transport. From this moment on, Austerlitz's inquiries, erratic until now, meant to help him forget, acquire a clear sense and are meant to find clues about his parents. He abandons nineteenth century history and fully enters the gloomy twentieth, with its intolerable years before World War II, the insecurity years, years of vital risk for the Jewish community.

In this second part of the novel, the character visits three European countries in which he hopes he might find the clues he is searching for: Czechoslovakia, Germany and France. In Prague, he finds the house in which he had lived before his exile and speaks with a neighbour who knew him as a child, Vera, and who, above all, knew his parents and their unfortunate fates. Thanks to Vera's testimony, Austerlitz knows that his mother was confined in a Nazi concentration camp in Germany and that she probably died there. In the same way, he discovers that his father escaped to France, shortly before the Germany's annexation of Czechoslovakia.

This is how the narrator meets Austerlitz in Paris. In his search for clues about his father, this city plays a fundamental role. The narrator recounts the three encounters he had with Austerlitz in the French capital, in a bar called La Havane, in the Avenue Auguste Blanqui, next to the La Glacière metro. During these encounters, Austerlitz details how he came to Paris looking for his father, Maximilian Austerlitz, who in a desperate attempt to escape from the growing threat Hitler represented, escaped to France. Apparently he went to Paris.

However, that was not the first time Jacques Austerlitz visited Paris: in the 1950s he had been there too, although for purely academic reasons. The comparison between the two moments reveals the profound metamorphosis undergone by the character and by the space in which he moves.

By the end of the 50s, still determined to ignore his past, devoted to deeply erudite reflections on the history of the architecture, Austerlitz dedicates his time in Paris to going to the National Library, then located in rue Richelieu. The relation between the character and the space is, at this point, at least paradoxical, since a space dedicated to memory – a video promoting the library is entitled *Tout la mémoire du monde* – is used for the oblivion. Austerlitz goes to the memory temple with the clear intention of forgetting:

> During my first stay in Paris [...] I tried not to let anything distract me from my studies. In the week I went daily to the Bibliothèque Nationale in the rue Richelieu, and usually remained in my place there until evening, in silent solidarity with the many others immersed in their intellectual labours, losing myself in the small print of the footnotes to the works I was reading, in the books I found mentioned in those notes, then in the footnotes to those books in their own turn, and so escaping from factual, scholarly accounts to the strangest of details, in a kind of continual regression expressed in the form of my own marginal remarks and glosses (363).

It is precisely there where he meets Marie de Verneuil, and with her he will experience what seems to be the only happy, though fragile, moment in the otherwise solemn life of this character. The transformation felt by the character is reflected in the space around him too. Austerlitz usually prefers, not only in Paris, spaces rather isolated, marginal, in the suburbs, semi-abandoned, or places like cafeterias and especially train stations: "non-spaces" for a "non-individual." During the time he is involved with Marie, on the contrary, Austerlitz takes walks in the Jardin de Luxembourg, in the Tuileries, the Jardin des Plantes or the Museum of Natural History, jumping from the periphery to downtown Paris.

A week without Marie, however, is enough for his stability to collapse. Every now and then, Marie visits her parents or relatives, and during those moments, which he confesses made him feel a terrible anxiety, Austerlitz goes back to his habitual marginal spaces: Montreuil, Malakoff, Charenton, Bobigny, etc. While returning from one of these walks, in which he had seen the horrible Veterinary Museum, located in Maisons-Alfort, Austerlitz begins to have episodes of passing out, more and more often after this first time. This fainting is accompanied by temporal yet complete loss of memory, which according to the psychiatric terminology, are the symptoms of hysterical epilepsy.

This is an episode previous to his nervous breakdown during the summer of 1992. For a long time in his adult life, the time in which he cultivates oblivion, Austerlitz suffers various episodes of fainting and loss of memory. At the structural level of the novel, they all announce an argumentative turn. This is the case when after the nervous breakdown, the character discovers the initial clues about his real history, while he visits an old bookstore next to the British Museum. After he faints for the first time in Paris, by the end of the 50s, Austerlitz will experience a moment of revelation, although much more confused and with less measurable consequences than in 1992.

Indeed, after his recovery, Marie and Austerlitz will take a walk through the semi-abandoned area between the Gare d'Austerlitz and the Quai d'Austerlitz, on the left bank of the Seine. There they witness, in an almost surreal episode, a circus show whose climax is a musical representation for the two of them only. The scene has the value of an epiphany for Austerlitz, who in turn does not know to explain the reasons for which the space and the music resound so powerfully inside him:

> I cannot say what it was that the five circus performers placed that Saturday afternoon in the circus tent beyond the Gare d'Austerlitz […] but it seemed to me, he added, as if the music came from somewhere very distant, from the East, I thought, from the Causasus or Turkey. Nor can I say what was suggested to my mind by the sounds produced by the players (382-383).

The story of Austerlitz's first time in Paris stops here and the narrator scarcely gets any more information. He knows that the protagonist's next destiny was London and that unfortunately, his affair with Marie did not work. Next, Austerlitz speaks about present-day Paris, in which he and the narrator find themselves. Austerlitz underlines the continuity of space, since that area in between the Gare d'Austerlitz and the Quai d'Austerlitz is where the new Bibliothèque Nationale was built.

Austerlitz is no longer that lost individual who went to the rue Richelieu in order to feed oblivion by investigating a past which was not his own. He is now a tenacious researcher of his past who would do anything to break the oblivion. Nevertheless, the space is not on his side anymore. The new library is not "a temple of memory", but a sinister place with a Kafkian appearance, in which everything seems to turn into an obstacle for the information search:

> The new library building, which in both its entire layout and its near-ludicrous internal regulation seeks to exclude the reader as a potential enemy, might be described [...] as the official manifestation of the increasingly importunate urge to break with everything which still has some living connection to the past (398).

Already, access to the library is narrated by Austerlitz as if it were an odyssey; it seems destined to dissuade those who want to enter; to humiliate them if they try it; to dehumanize them, if they finally make it:

> When I first stood in the promenade deck of the new Bibliothèque Nationale, said Austerlitz, it took me a little while to find the place where the visitor is carried down on a conveyor belt to what appears to be a basement storey but, in reality, is the ground floor. This downwards journey, when you have just laboriously ascended to the plateau, struck me as an utter absurdity, something that must have been devised [...] on purpose to instil a sense of insecurity and humiliation in the poor readers (389).

The difficult access to the library does not end here. Once inside, the reader must explain at an information point in the hall what his requests are and if they are only slightly different from the expected, the visitor is asked to go to a separate cubicle, where the visitor is asked for more exact data "as if you were on business of an extremely dubious nature, or at least had to be dealt with away from the public gaze" (390).

Not only for its "layout" or its "internal regulation," but for its topographic position, the library is an obstacle for the access to the past. As Austerlitz explains, the ground where the library was built, the same space where decades before Marie and he attended the strange circus show, belonged to the French Jews. It was expropriated and turned into a warehouse where the Germans kept the loot taken form the homes of the Jews of Paris: "The whole affair is buried in the most literal sense beneath the foundations of our pharaonic President's Grande Bibliothèque"[11] (403).

In Paris's spaces, the same outline that guides the entire novel is reproduced. Before the transformation occurred in the character in 1992, he has no connection with the space around him; it, space, can be mysteriously meaningful from time to time, or a symbol of the character's own marginality. Starting with his search for identity, however, space becomes an obstacle to encountering information. Time adds new layers to those already existing, in a process similar to what oblivion does to memory. History is, thus, an obstacle to history. The metamorphosis of the Bibliothèque Nationale is the symbol of the dehumanization of knowledge, of an increasing difficulty of accessing one's own past, of the burying of twentieth century Europe's roots into political grandeloquence. Here, too, a fatalist view of the passing of time, of the transformation of emblematic spaces, on a personal and collective level.

Dark times here and there

What can a profound Romantic novel like Victor Hugo's *Notre-Dame de Paris* have in common with a postmodern novel like Wolfgang Sebald's *Austerlitz*? This was the question in the beginning. And it was said that Paris becomes in both literary works a symbol for the metamorphosis of places.

Hugo's work is centred in Paris, more specifically in the Cathedral of Our Lady; indeed, some critics have argued that this building is the protagonist of the novel, more than the characters presented in it. In *Austerlitz*, the narrator meets the main character, Jacques Austerlitz, in different cities in Europe, often in representative buildings (mainly train stations), and their conversation is very often centred on buildings. Paris is here only one of the evoked cities, but an especially relevant one: in the middle of the novel we learn Austerlitz is actually seeking his father, lost

[11] In the novel, Austerlitz repeatedly recounts the storage of various buildings and specifically what can be found in their foundations, often sinister places. In the excavations done in 1984 in the train station of Broad Street, for example, an area Austerlitz strolled about by that time, were found more than four hundred corpses. In the present of the story, neither the previous district nor the train station exist any more (185-86).

in the confusion of World War II. The last trace seems to lead him even to Austerlitz train station in Paris, where his father probably has taken a train to escape. Is in Paris where the last encounters, the last conversations with the narrator take place, in this novel on which these encounters are the essential points.

In both novels, Paris plays a key role. And in both novels a metamorphosis of a place is not only described, but deplored. In Hugo's *Notre-Dame de Paris* a whole chapter is devoted to the cathedral, to the monument located in the middle of a labyrinth of small streets. But the cathedral is only one example – the most paradigmatic one – of many Hugo describes in the same manner, in the comparison between the fifteenth century and his own time. The perspective from which Hugo looks to this passing of time is apparently a nostalgic one: he deplores the transformations of the church – and of the other buildings – during the centuries, sometimes due to pacific modifications made by architects under the sway of new styles, sometimes made violently, during the Revolution, in which the cathedral was transformed into a temple of Reason. In any case, it is a fatalistic view: the transformations seem to be inevitable; they can be observed, they can be described, they can be deplored. But they are rooted in a general law of the history of societies.

Notre-Dame de Paris was published in 1831. Approximately one century before, in 1721, the first collections of France's National Library were transferred to the "Site Richelieu," the now "old" seat of the Library. In Victor Hugo's time, this "Bibliothèque Nationale" is amplified and reorganised. This is the library Austerlitz compares with the new one, built in 1988 under the auspices of François Mitterrand. In these years of Mitterrand–fever for Paris, in which he was a kind of *super–Major* of the city, in the years of enthusiasm for cultural infrastructures (museums, concert halls, etc.), a modern, high-tech library was created. But in Austerlitz's eyes it is an inhuman monster which provokes nostalgic remembering of the old reading rooms[12].

After that, the second part of the nineteenth century, the time of industrialisation – which plays a considerable role in the novel – is

[12] "The old library in the rue Richelieu has been closed [...] said Austerlitz, the domed hall with its green porcelain lampshades which cast such a soothing, pleasant light is deserted, the books have been taken off the shelves, and the readers, who once sat at the desks numbered with little enamel plates, in close contact with their neighbours and in silent harmony with those who had gone before them, might have vanished from the face of the earth" (Sebald 385-86).

described as dark[13], and the twentieth century as the time in which the rootlessness and identity crisis appeared. Now the cathedral has been substituted by a library, in a typical metamorphosis of the *sacrum* by post–modernity. But strangely enough, the reaction remains the same: the transformation of space is deplored and used as an instrument for criticising one's own time as inhuman and barbaric. With one small difference: what Hugo considered a dark time (his own time) is now seen as the time in which a human sense was still possible. For Victor Hugo, his time had destroyed harmony; for Austerlitz, Hugo's time maintains a human dimension.

Is this only a change of perspective? Or is it a consequence of the fact that the first half of the twentieth century has been indeed a great European disaster? The transformation of place is in both cases a symbol of the interpretation of history, a gloomy interpretation. In Austerlitz's view this vision sounds more convincing, because it is exposed only by telling a story, his own story, whereas Victor Hugo explicitly includes the transformation into an exhaustive theory about the societies' evolution. But in the end, both views contemplate the transformation of places with a sceptical attitude, as a loss of something essential.

Works Cited

Hugo, Victor. *Notre-Dame de Paris: Les Travailleurs de la mer*. Paris: Gallimard, 1992.

Torrego Salcedo, Alberto. Introducción. In *Notre-Dame de París*. Madrid: Gredos, 2006.

Sebald, W.G. *Austerlitz*. New York: Random House, 2001.

---

[13] "Austerlitz launched into a discourse of over two hours on the way in which, during the 19th century, the vision of model towns for workers entertained by philanthropic entrepreneurs had inadvertently changed into the practice of accommodating them into barracks – just as our best-laid plans […] always turn into the exact opposite when they are put into practice" (Sebald 37-38).

DICKENS'S *DAVID COPPERFIELD*: WORLDS, LANDSCAPES AND NARRATIVE

ZELMA CATALAN

Abstract

The paper analyses the transformations of the spatial perspectives in the ontological landscapes of the eponymous hero of Dickens's *David Copperfield*. It argues that the concept of place in this work and in the Bildungsroman in general has a relational meaning. Places gather significance by virtue of their mutual spatial disposition and of the dynamics of change. The childhood scenes are set in a landscape where the narrator's self is fully integrated and where the sacred domain seems to be continuous to the profane one along a horizontal axis. The Wickfields' home and Agnes in particular introduce the vertical axis which allows true access to the transcendental, but it is barred by the presence of Uriah Heep and Steerforth in David's life. Only after they have been eliminated and he has had his epiphany can the narrative affect the true union of interiority and exteriority and reach closure.

Between his first glimpse "into the blanks of [his] infancy" (13) and the closing look "back – once more – for the last time" (865), the narrative of Charles Dickens's *David Copperfield* covers all the key elements of the physical, cognitive and psychological journey typical of the Bildungsroman narrative. Starting as an infant in Suffolk, he ends up living in London happily married and with a successful career as a professional writer. In most studies of the genre, these two places – the province and the city – have been regarded as the locales of the two states between which the Bildungsroman narratizes its social and psychological teleology. As Franco Moretti has pointed out, the Bildungsroman asserts the possibility of the socialization of the individual through the fusion of "external compulsion and internal impulses into a new unity until the former is no longer distinguishable from the latter" (16). Unlike Georg Lukacs, from whose *Theory of the Novel* his influential study takes its theoretical cue, Moretti emphasizes that the synthesis aesthetically confirmed by the Bildungsroman is one between two complementary but essentially conflicting ideologies of self and society (17). Developing Moretti's point further, one can find in it the source of what both Lukacs (142) and Buckley in his now classical study of the English Bildungsroman (23-24) see as the greatest difficulty for all writers working in the genre: to give their works an organic ending. Indeed, the aesthetic compromise required for a "happy ending," is predicated on the fact that modernity, metonymically represented by the city, can never offer all that the hero needs for a fulfilled interior life.[1] The problem is especially visible in the nineteenth-century English novel, of which *David Copperfield* is a notable example. Until the mid-century, it continues rather than severs the relationship between man and nature set up by Romanticism. The narrative ending therefore requires that some of the ideological tenets of the beginning remain, albeit in a subdued or modified form, that the final landscape of mundane reality embraces the initial landscape of desire for the transcendent. The present paper aims to analyze Dickens's effort in *David Copperfield* to conjoin these two landscapes through modifications in their spatial configuration and by using space and place as more than a setting of the action. My argument is that the concept of place in this work and in the Bildungsroman in general has a relational meaning. Like the country and the city, places, both in descriptive and narrative sections of the text, gather significance by virtue of their mutual spatial disposition and of its dynamics of change in the context of the characters' ontological landscapes.

By "ontological landscape" I mean the abstract analogy of a world model belonging to an individual or to a society (Pavel 137-143). In the nineteenth-century novel, as noted earlier, the architectonics of this landscape comprises the sacred world, as well as the variety of profane worlds – of work and leisure, of emotional and practical experience. The concept of "world" itself finds its literary-theoretical explanation and analysis in the mapping of the theory of possible worlds onto contemporary narratology. In her influential series of papers, Marie-Laure Ryan treats the universe of narrative fiction as comprising the fictional Actual world and worlds relative to the character – Knowledge-worlds and

---

[1] Bakhtin is very much aware of this when he distinguishes between the Bildungsroman proper and the novel of emergence. In the latter, he says, the orientation is towards the historical future, "a completely new, *spatial* sphere of historical existence" (24).

their hypothetical extensions, Model-worlds including Wish-worlds and Moral values-worlds; these worlds are modally differentiated (Ryan 720-732). Movement or transformation within and among the narrative worlds creates conflicts in the narrative universe. Out of the productive ones among them – that is, the ones "[whose] experiencer is in a position and is willing to take steps towards [their] resolution" (Ryan 733) – emerges the plot as a unity of narrative events. Bringing together Ryan's and Pavel's treatment of fictional worlds, I would suggest that a productive conflict can be defined in spatial terms: it is the consequence of a disruption in the arrangement of worlds within the ontological landscape. In *David Copperfield*, however, each disruption entails not only a change within the landscape but within the respective worlds as well, signalling an external, as well as an internal conflict.

The first part of Chapter Two of *David Copperfield*, "I Observe," presents the ontological landscape of the narrator's early childhood as a homogeneous whole. The Knowledge-world of his young self fully coincides with his Model-world, delineating a space where everything is as it should be and the limits of cognitive and emotional experience are the limits of desire. One might expect that the syntactically ambiguous title "I Observe" would lead to a random collection of objects, people and scenes that make up the notion of "memory." But in a novel that so strongly promotes the moral significance of self-discipline, even the earliest remembrances are subjected to a strict ordering and to a common principle of inner structuring. David's memory is indeed well landscaped, and each element in it takes its significance according to its relative position to the other elements and to David himself – both of the time he is recalling and of the present moment of writing.[2] His series of recollections is therefore chronologically constructed so as to mark stages in his biological growth from infancy to childhood but also to prepare the causal ground for the events that are to follow.

From the very beginning, however, the sense of temporal continuity relies also on spatial relativity, which represents the structural backbone of each scene in the series. Thus, in the narrator's earliest distinct memory, obviously of the time when he was a toddler, his mother and Peggotty emerge as two material entities "*at a little distance apart, dwarfed to my sight* by stooping down or kneeling on the floor, and I going unsteadily

[2] In his "Cookery, not Rookery: Family and Class in *David Copperfield*" Chris Vanden Bossche makes the similar comment that even during his holiday, when David longingly remembers the days of his childhood, "this former life is already a creative memory, the product of his resistance to Murdstone and Creakle" (33).

from one to the other" (13, italics added). In the next scene, he is already allowed to move on his own and explore the house and the backyard, with its "quantity of fowls that *look terribly tall to me*, walking about in a menacing and ferocious manner" (14, italics added). The scene in the church that follows belongs to an even later period, as David is now able to read "the monumental tablets on the wall" (15). Here, the spatial perspective charts the direction of his glance which oscillates rhythmically between people and objects within the church and outside, just as each of his urges to commit some misdemeanour is followed by the certainty of being appropriately punished. And in the last scene in the series, which takes place in the parlour of Blunderstone Rookery, he is already capable of even more mature judgements that again superimpose a cognitive value onto relative position in space: watching his mother taking a moment of rest in her armchair and unselfconsciously "winding her bright curls round her fingers, and straightening her waist," he says that "nobody knows better than I do that she likes to look so well, and is proud of being so pretty" (16).

All of these early scenes are set in a physical landscape characterized by a horizontally built topography, where every element is clearly visible and comprehensible and where there are no obstacles to impede the child's movements and gaze. Even the borders between inside and outside are permeable: the doors and windows allow free access for motion and observation, as, for instance, during the Sunday service: "I look at the sunlight coming in at the open door through the porch, and there I see a stray sheep – I don't mean a sinner, but mutton-half making up his mind to come into the church" (15). Or in the next recollected childhood vision: "And now I see the outside of our house, with the latticed bedroom-windows standing open to let in the sweet-smelling air, and the ragged old rooks'-nests still dangling in the elm-trees at the bottom of the front garden" (ibid.). Blunderstone Rookery and its environs represent a world in which the child's "I" is fully integrated and where neither his ocular vision nor his cognitive and imaginative endeavour is hampered by anything beyond the limits of his interpretation.

In an indirect manner, however, the style of the initial section of Chapter 1 makes more mature ontological commitments. It intimates the existence of another, sacred domain continuous to but beyond that delineated by the child's sensory perceptions and cognitive potential. In the two "outside" scenes, the garden and the churchyard appear as a natural extension to the happy domestic interior, a place for play and enjoyment. Still, the language in which they are presented imparts to the exterior the status of a prelapsarian Edenic world. The distancing of the sacred from the profane and the superimposition of a more complex ontological landscape onto the actual takes place in two stylistically foregrounded sentences. "There is nothing half so green that I know anywhere, as the grass of that churchyard; nothing half so shady as its trees; nothing half so quiet as its tombstones" (14), and later,

> Now I am in the garden at the back, beyond the yard where the empty pigeon-house and dog-kennel are – a very preserve of butterflies, as I remember it, with a high fence, a gate and a padlock; where the fruit clusters on the trees, riper and richer than any fruit has been ever since, in any other garden, where my mother gathers some in a basket, while I stand by, bolting furtive gooseberries, and trying to look unmoved (16).

Critic John McGowan aptly notes that the former sentence "suggests the limits of a reliance on the purely visual apprehension of reality" (4); the same can be said of the other one, too. From the measured parallelism based on the use of the comparative, there emerges an added meaning pointing to the absolute value of the objects and their features – a uniqueness that grants the pastoral atmosphere a life beyond time and beyond the scope of the individual momentary experience. It remodels the child's world in a fashion that cannot be comprehended within his Knowledge-world but is determined by considerations which anticipate the ontological divisions that are later to shape the Bildungsroman plot.

It is, in fact, in this kind of ontological landscaping that the first productive conflict in *David Copperfield* originates. For the appearance of the Murdstones totally rearranges David's world, literally as well as metaphorically. Nothing is "like itself"(44): David is not only removed to a cold and dismal room but the whole interior of the house now becomes a place of psychological and even physical torture. The outside, on its part, acquires its own active and controlling centre: the formerly empty kennel is now "filled up with a great dog – deep-mouthed and black-haired like [Murdstone]" (44) which stops the boy from his customary rambles in the yard. More than a physical obstacle to the child's free movement, the dog starts an imagistic series that will find its dramatic extension in the subsequent enforced animalization of David himself. This process of loss of human identity begins with his biting Murdstone's hand, continues in the school where he is made to wear a placard that reads "Take care of him. He bites" and culminates in his dehumanization at the warehouse. Although, as he states, he never sinks to the level of the other common boys and men at Murdstone and Grinby, he lives a life precariously poised between physical and emotional survival and death.

With his days passing in silent misery and incessant hunger, he is little better than a stray animal: "I know that I worked, from morning until night, with common men and boys, a shabby child. I know that I lounged about the streets, insufficiently and unsatisfactorily fed. I know that, but for the mercy of God, I might easily have been, for any care that was taken of me, a little robber or a little vagabond" (162).

The whole section of the narrative from Murdstone's arrival on the scene up to David's escape from Murdstone and Grinby is dominated by references to borders, limits and impeded or prohibited freedom of movement between places. It is noteworthy that from the moment of his first appearance until David's holiday at home, Murdstone is presented in an upright position, as is his sister. Their tall figures and their linguistic metonyms – the peremptory orders they issue – act as barriers between the boy and his mother. And when they forbid him to associate with Peggotty (119), his access to the kitchen – another domain of happiness and love – is similarly barred. It is only logical therefore that David's actions as he reaches the stage of independent decisions become largely governed by his desire to recover the undivided landscape of his childhood world and to glide uninhibited from domestic idyll to a nature paradise and back. His singleminded effort to marry Dora proceeds from this compulsive longing for a return to the Model-world of his childhood, where adjacent to the domain of material reality lies an elusive sacred centre. Significantly, just before he sees Dora for the first time and promptly falls in love with her, he passes through the garden of her father's house. As if to revive the image of the Earth-Eden continuum is the brief description of that garden, which takes the same stylistic form as that of the one in his childhood home: "There was a charming lawn, there were clusters of trees, and there were perspective walks that I could just distinguish in the dark, arched over with trellis-work, on which shrubs and flowers grew in the growing season" (384). And in the scene in the greenhouse from the following morning, the superlative forms grant Dora the status of absolute perfection, matching her, as all critics have recognized, with the image of David's mother: " I never saw such curls – how could I, for there never were such curls! – as those she shook out to hide her blushes" (390).

The ontological planning that David applies to his Knowledge-world in this period is, however, an error which he, as a Bildungsroman hero, must realize and correct. His short unhappy married life is in fact his period of what Pavel terms "ontological stress" (142), of his struggle to conceptualize and adapt to a formerly inexistent distinction between wish and actuality. But if this state of confusion fails to turn into a productive conflict, it is because he does not recognize that the landscape of the Actual world does contain a sacred centre. It is not, however, to be found as a horizontal projection of the profane. Indeed, the reason why David is unable to perceive it is that he continues to arrange his Model-world horizontally, the same way as he did during his childhood. Yet the new perspective has been available to him from much earlier, from the moment he entered the Wickfields' home in Canterbury. There, with Agnes's first appearance from her room upstairs and her later emblematic posture with her finger pointing upwards, the physical and ontological space undergoes a restructuring along the vertical dimension. And although the references to verticality remain for a long time indirect and do not surface in the text, they gradually coalesce into an entity that is of decisive importance for the plot movement and its resolution. For this entity encompasses the three people whose presence defines David's life before, during and after Dora's entry into it. They are Uriah Heep, Steerforth, and, of course, Agnes herself.

The Wickfields' home is divided into the upper region inhabited by Agnes, and the lower region governed by Uriah Heep. Not only his locatedness in the nether quarters but Heep's very appearance takes its full meaning from this opposition. The repeated references to his ghostly and ghastly figure recreate the house as, in the words of Gaston Bachelard, one "with cosmic roots" (23) reaching into the diabolical. His abode in the small secluded room downstairs turns it into a space of "an anthropo-cosmic fear" of the uncanny (ibid.). While the realm of death, symbolized by Mr. Copperfield the elder's grave, was at the beginning implicitly defined as a positively-charged locus of the sacred, now David's recurring references to the clammy touch of Heep's hands produce its reverse, negatively-charged experience – the physical sensation of contact with Hell, a primitive terror that governs every scene where Heep is present.[3]

No doubt David's revulsion from Uriah Heep has to do with the parallels with himself that his very narrative brings out, as well as with the need to conceal or manipulate his own response. Heep's place in the physical and ontological landscape, however, is no less important to the narrative development of the novel. For Heep has the power to invade the whole space and paralyze all movement in it. Not only does he literally ascend in status by taking over David's old room upstairs but, as Vincent

[3] Simon Edwards develops the issue in his "David Copperfield: The Decomposing Self" along the lines of David's and Uriah's common guilt of masturbation (350). Vincent Newey offers the same treatment but in greater detail (129-34).

Newey points out (138), whenever David can and should take action to prevent his dirty dealings, he finds himself incapable of this. Unlike his Biblical namesake, the utmost he can do is strike Uriah's cheek on the night after the latter has implicated him in the disruption of the Strongs' domestic peace. But even then David feels no pride in his action, "only less mean than he. He knew me better than I knew myself" (613). Like Blake's Serpent, Heep insinuates himself at the very centre of what David has regarded to be a temple to his earthly Madonna – Agnes. Uriah Heep inverts the principle of coherence and gives a new meaning to the domain of the sacred: he infuses it with the power to degrade and destroy. For once, Micawber's florid language matches reality such as David himself sees it: Uriah Heep is indeed "transcendent and immortal" (702). Still, because he is the embodiment of a kind of primordial fear, he can only be routed and purged by an equally irrational force – by the exorcising energy of humour epitomized and expressed by Mr. Micawber. Having invaded the whole of the landscape and thus collapsed the distinction between its worlds, Uriah Heep makes any productive conflict impossible. That is why Micawber's intervention is so crucial: without it both the life and the narrative of David Copperfield would be stuck.

No such exorcism is, however, possible with the other source of narrative subversion in *David Copperfield* – Steerforth. Heep radicalizes evil by sending it "down" into the realm of mythically subconscious. Nevertheless, though indefeasible by any rational or logical means, Heep is at least intelligible, for the semantics given to him by the descriptive language and by the narrative is not only familiar but also stable and coherent. Unlike him, Steerforth relativizes the distinctions between moral values, while at the same time holding an irresistible appeal for David. David's homoerotic fancy for Steerforth is in fact the greatest obstacle he must overcome. In the traditional Bildungsroman plot the corrupting influence of modern city life is epitomized by the stereotypical femme fatale and the protagonist's rejection of her seductive power is the major narrative event that enables the final fusion of interiority and exteriority. In this novel, however, Dora hardly fits that image: instead, its function is taken up by Steerforth. Again, the semantics around Steerforth that refers to space and location is telling. On the whole, his presence is marked by an insistence on horizontality. His name itself accentuates horizontal movement, dramatized in his journeys to and from Yarmouth, later on around the Continent with Emily and finally in his fatal return to England on the schooner. Even the static image in which David seals his remembrance of Steerforth is dominated by the horizontal dimension. As Vincent Newey notes, "Steerforth's characteristic symbolic pose is the sinuous and horizontal line, sleeping head on outstretched arm, passive, aesthetic, alluring, self-justifying" (124). The negative charge he gives to horizontality carries a narrative energy that disambiguates retrospectively even Emily, hallowed by David's childhood remembrances but oscillating between the positions of self-made and innocent victim. Thus, during the very first day David and she spent together as little children, she "started from my side, and ran along a jagged timber which protruded from the place we stood upon, and overhung the deep water at some height, without the least defense" (36). This causes the narrator to remark with hindsight that it would have been better for her to have actually completed that journey to death (37). What the spatial features of Steerforth's presence and appearance seem to assert therefore is that reaching a green, shady, quiet place of eternal rest, either in happy domesticity or in peaceful death, is impossible through movement enacted on the horizontal plane, as David had initially imagined.

David's sacralization of Steerforth in the face of his friend's very earthly and morally inexcusable traits becomes a problem impeding narrative progression. Since David creates around him a Wish-world which competes with Agnes's in the novel's teleology, Steerforth stands as Agnes's rival in the narrator's ontological landscape. To achieve the necessary rearrangement of his worlds, David needs to clear the space from Steerforth and leave Agnes as the sole occupant of the sacred domain. Yet, Steerforth is so deeply entrenched in David's structure of valorizations that he cannot be dealt with merely through narrative contrivance, as happens with Dora, or through humour, as is the case with Uriah Heep. This task needs a superhuman agent and is indeed performed through quasi-divine intervention in two major narrative events. One is the storm in which Steerforth dies and the other is David's epiphanic experience in the Alps. In both, the language enacts the necessary rearrangement of space, so that the vertical orientation of David's world-model can finally solidify into a narrative perspective sufficiently prominent to enforce the reorientation of the narrative towards the desired end.

In the description of the storm, the spatial reconfiguration takes shape in the course of a recurring pattern of fragmentation, achieved by means of uneven, broken syntax but, more importantly, through nominal and adverbial use which sends reference into all points of the compass. The verbs, too, are mostly of movement up or down, as in the paragraph which starts the long account:

> It was a murky confusion – here and there blotted with a colour like the colour of the smoke from damp fuel – of flying clouds, *tossed up* into most remarkable heaps, suggesting greater *heights* in the clouds than there were *depths below* them to the *bottom* of the *deepest hollows* in the earth, through which the wild moon seemed to *plunge headlong*, as if, in a dread disturbance of the laws of nature, she had lost her way and were frightened. There had been a wind all day; and it was *rising* then, with an extraordinary great sound. In another hour it had much increased, and the sky was more *overcast*, and blew hard (778, italics added).

Or in the following, which employs the whole repertoire of verbs, prepositions and adverbs signalling division, detachment and removal:

> We came to Ipswich-very late, having had to *fight* every inch of ground since we were ten miles *out of* London; and found a cluster of people in the market-place, who had *risen from* their beds in the night, fearful of *falling* chimneys. Some of these, congregating *about* the inn-yard while we changed horses, told us of great sheets of lead having been *ripped off* a high church-tower, and *flung into* a by-street, which they then *blocked up*. Others had to tell of country people, coming in from neighbouring villages, who had seen great trees lying *torn out of* the earth, and whole ricks *scattered about* the roads and fields. Still, there was no abatement in the storm, but it blew harder (ibid.).

Throughout that part of Chapter 55, the semantic consistency works together with the measured crescendo of emotional intensity to evoke the sense of an impending catastrophe that only seemingly takes place in a particular physical locality. The "dread disturbance of the laws of nature" from the quotation above, the "rending and upheaving of nature" (780), exert their cataclysmical effect on the whole of David's representation of the world and erase its spatial certainties. As Robert Lougy points out, "[t]he storm that breaks over Yarmouth and the ocean that threatens to pull the structures of the town back into itself represent the potential triumph of formlessness over form, of death over life" (53). That Steerforth's death turns out to be an event of such a shattering impact on David is largely due to the gap that the storm, both as a physical and as a linguistic occurrence, has opened in David's representation of reality and to the related fact that it has failed to impose the necessary new order: "all the world seemed death and silence, broken only by his mother's moaning" (793).

David's dejection subsequent to his emotional losses stems from his realization that he must accommodate himself to a newly configured landscape but that he is unable to do so. Again, the semantics of the language with which he describes his inner state points to desolation and emptiness. What he can only see is "a ruined blank and waste, lying dark around me, unbroken to the dark horizon" (806). Yet, this is still the view of someone looking only horizontally. It is not until in the Swiss valley he turns his gaze upwards that he feels the spark of life kindled again:

> I came into the valley, as the evening sun was shining on the remote heights of snow, that closed it in, like eternal clouds. The bases of the mountains forming the gorge in which the little village lay, were richly green; and high above this gentler vegetation, grew forests of dark fir, cleaving the wintry snow-drift, wedge-like, and stemming the avalanche. Above these, were range upon range of craggy steeps, grey rock, bright ice, and smooth verdure-specks of pasture, all gradually blending with the crowning snow. Dotted here and there on the mountain's-side, each tiny dot a home, were lonely wooden cottages, so dwarfed by the towering heights that they appeared too small for toys (807).

In a High-Romantic fashion,[4] David experiences an epiphany that results in the imposition of inner and outer stability in his world. The unearthly music he hears, the voice of "great Nature" coming from a space above (807) opens up that dimension which is the homeland of the transcendental and where Agnes's image, with her finger pointing upwards, has been beckoning him all along. As he comes out of her house after his three-year absence, he describes his feelings in the same religious language that marks her textual presence throughout the book. Now, however, it carries the enriched modulation of a love beyond this life, an emotion nonetheless precious for being even stronger in heaven than it was in this world: "thinking of her, pointing upward, thought of her as pointing to that sky above me, where, in the mystery to come, I might yet love her with a love unknown on earth, and tell her what the strife had been within me when I loved her here" (837). The circle is complete, with death again reestablished as a possible site of happiness as at the beginning of his narrative, but this time it is placed in a comprehensible and ontologically stable landscape dominated by the sacredness of a centre above the mundane, transitory, and profane.

David Copperfield's ending – the happiest among all Dickens's novels – may seem too pat, too contrived, sheer wish-fulfillment.[5] It is,

---

[4] Parallels between Dickens's romantic poetic in this section have extended to Wordsworth's *Prelude* and Byron's *Childe Harolde* (Newey 151-153), Carlyle's *Sartor Resartus* and Tennyson's *In Memoriam* (Buckley 40-41).

[5] James Kincaid in "Viewing and Blurring in Dickens" speaks for most critics when he claims that in the final account David capitulates to circumstances rather

however, the logical conclusion to a dynamic narrative of ontological transformation which starts from the sacralization of the pastoral and ends in a sacralization of the domestic. And this conclusion encompasses the spatial transformation of the landscapes into which David's world representations are shaped: Agnes's face, "shining *on* me like a Heavenly light by which I see all other objects, is *above* them and *beyond* them all," "I turn my head, and see it, in its beautiful serenity, *beside* me," "O Agnes, O my soul, so may thy face be *by* me when I close my life indeed; so may I, when realities are melting from me, like the shadows which I now dismiss, still find thee *near* me, pointing *upward*! (869, italics added).

Throughout his journey, David Copperfield has been a perfect embodiment of Lukacs's Bildungsroman hero: a soul carrying "as a sign of its tenuous, but not yet severed link with the transcendental order, a longing for an earthly home which may correspond to its ideal – an ideal which eludes positive definition but is clear enough in negative terms" (132). Dickens's *David Copperfield* is the story of such a soul through which the novelist, however controversially, attempts to positively establish the terms of its ideal world. That it is a unique and unsustainable experiment is proved by his second, and very different, Bildungsroman, *Great Expectations*.

Works Cited

Bakhtin, M. M. "The *Bildungsroman* and Its Significance in the History of Realism (Toward a Historical Typology of the Novel), in M. M. Bakhtin, *Speech Genres and Other Late Essays,* ed. Caryl Emerson and Michael Holquist, trans. Vern W. McGee. Austin: University of Texas Press, 1986.

Bloom, Harold. *Charles Dickens's* David Copperfield. Philadelphia: Chelsea House Publishers, 1987.

Buckley, Jerome Hamilton. *Season of Youth*. Harvard: Cambridge UP, 1975.

Bachelard, Gaston. *The Poetics of Space*. Boston: Beacon Press, 1994.

Dickens, Charles. *David Copperfield.* London: Pan Books, 1967.

Edwards, Simon. "David Copperfield: The Decomposing Self." *The Centennial Review*, 29.3 (1985): 328-52.

---

than triumphing over them (107). A similar and better argued opinion can be found in Lougy's *Remembrances of Death Past and Future: A Reading of* David Copperfield.

Kincaid, James R. "Viewing and Blurring in Dickens: The Misrepresentation of Representation." *Dickens Studies Annual* 16 (1987): 95-111.

Lougy, Robert E. *Remembrances of Death Past and Future: A Reading of* David Copperfield. *Dickens Studies Annual* 6 (1977): 72-101. Repr. in Bloom, H., 47-66.

Lukacs, Georg. *The Theory of the Novel*. London: Merlin, 1988.

McGowan, John P. "David Copperfield: The Trial of Realism." *Nineteenth-Century Fiction*, 34.1 (1979): 1-19.

Moretti, Franco. *The Way of the World: The Bildungsroman in European Culture*. London: Verso, 2000.

Newey, Vincent. *The Sciptures of Charles Dickens*. Aldershot: Ashgate, 2004.

Pavel, Thomas. *Fictional Worlds*. Cambridge: Harvard UP, 1986.

Peck, John (Ed.). *David Copperfield and Hard Times*. New York: St. Martin's Press, 1995.

Ryan, Marie-Laure. "The Modal Structure of Narrative Universes." *Poetics Today* 6 (1985): 717- 56.

Vanden Bossche, Chris. "Cookery, not Rookery: Family and Class in *David Copperfield*." *Dickens Studies Annual* (1986): 94-98. Repr. in Peck, 31-57.

CARIBBEAN METAMORPHOSIS: DE-ESSENTIALIZED COLONIAL IDENTITIES IN JEAN RHYS'S *WIDE SARGASSO SEA*

NAGIHAN HALILOGLU

Abstract

The metamorphosis I focus on in this paper is the transformation of certain characters from the cultural archive to signify differently in texts written in the postcolonial era. I argue that in *Wide Sargasso Sea* together with the character of Antoinette, the character of Rochester is also reconsidered, and certain experiences in the Caribbean, which remain silenced in *Jane Eyre* are made available to the reader. My investigation reveals a certain continuation in the white identities in the colonies, and reveals to us the metamorphosis of the white colonial subject in the space of the Caribbean, from colonizer to the depraved white colonial who is now herself open to exploitation like the natives of the island, or the indentured labour brought from Africa. Thus, *Wide Sargasso Sea* stages the transformations in colonial positions, and points to the ultimate porousness of categories imposed by hegemonic structures.

> I lived with that woman upstairs four years, and before that time she has tried me indeed: her character ripened and developed with frightful rapidity; her vices sprang up fast and rank [...] Bertha Mason, the true daughter of an infamous mother, dragged me through all the hideous and degrading agonies which must attend a man bound to a wife at once intemperate and unchaste (Brontë 303-04).

As a prequel to *Jane Eyre*, *Wide Sargasso Sea* is one of the most important works that deal with metamorphosis from the postcolonial cannon, due to its portrayal of the character of Antoinette, in whom readers are led to recognize Bertha from the Charlotte Bronte novel. In *Jane Eyre*, she is the mad wife in the attic who spoils Jane's chances to marry Rochester, the owner of the manor house she works at as a governess. However, at the end, Antoinette decides to set fire to the house in one of her frenzies, and thus makes Rochester a free man. In *Wide Sargasso Sea*, however, "the true daughter of an infamous mother" (Brontë 304). gets to tell her own story, and not only of her growing up years, but also of her marriage to Rochester, thereby revealing the hidden history of Rochester as well, a man who has made his money in the Caribbean as prototypical Victorian colonizer. Thus, in *Wide Sargasso Sea* both Antoinette and Rochester hatch into things that are already inscribed in *Jane Eyre*, but it takes the revelatory power of self-narrative of the characters in question to bring these metamorphoses about.

In this paper I aim to look at the ways in which narrative can function as a transforming or correcting force in inscribing literary characters in the cultural archive in the context of *Wide Sargasso Sea*. In order to do that, I will first look at the narrative form of the novel to gauge the possibilities it opens for re-representation of certain characters. I argue that the *Wide Sargasso Sea* is written in the "mad-witch" narrative persona under which certain, especially female writing from the Caribbean can be grouped. As an eclectic form, the mad-witch narrative, which I will talk about shortly, allows for the articulation of various subjectivities in the same narrative space. The "I" narratives, or self-narratives we encounter in *Wide Sargasso Sea* are symptomatic of the mad-witch model, which blurs the boundaries between subjectivities. Through this form hidden memories of both Antoinette and Rochester come out, revealing the similarities between the experiences of these two different colonial subjects, leading to the transformation of these characters. The transformation of characters the readers know from the Victorian canon, I argue, points to a wider mechanism of transformation that takes place in the colonial space - that is the blurring of the lines between the colonizer and the colonized. This transformation will be analyzed by an investigation of the way that intersubjectivity is staged in the novel and will reveal the intimacies occasioned by the resemblance of the experiences and memories of the two subjects.

Wide Sargasso Sea is the story of Antoinette, a white creole from the West Indies who marries an Englishman, and is, after a short and disastrous period of married life, shipped off to England in pretty much an incapacitated state. Although Rhys falls short of naming the husband Rochester, all the other names are consistent and the characters are recognizable from Charlotte Bronte's *Jane Eyre*. The figures of Antoinette and Rochester, in their embodiments in *Jane Eyre*, have come to stand for certain archetypes, and their stories, I argue, have become part and parcel of cultural memory when it comes to the colonial past. In their seminal work, tracing the transformation in the way literary criticism has conceived of the mentally disturbed heroines of the Victorian period, The *Madwoman in the Attic* Sandra M. Gilbert and Susan Gubar have read Antoinette, or Bertha Mason, as representing the suppressed psyche and sexuality of *Jane Eyre*:

> Thornfield's attic soon becomes a complex focal point where Jane's own rationality (What she has learned from Miss Temple) and her irrationality (her "hunger, rebellion and rage") intersect. She never, for instance, articulates her rational desire for liberty so well as when she stands on the battlements of Thornfield, looking out over the world (Gilbert and Gubar 348, 1979).

It is as if the madwoman has been disembodied, transported from the Real and entered the Symbolic in the Lacanian sense. However, Rhys transforms the character of Antoinette, not by reading her as an extension of *Jane Eyre*'s psyche, but by giving her a voice.

Wide Sargasso Sea is a narrative that has been divided into three parts of unequal length. The first narrative is the first person account of Antoinette's childhood. The second part starts with the narration of Rochester who speaks about his arrival to the island and his marriage, however, the narrating voice switches back and forth between him and Antoinette. The third part is narrated by one of the servants in the house that Antoinette is imprisoned in and the names of other personnel mentioned reveal this place to be none other than the Thornfield Hall of *Jane Eyre*. It is these changes in narrative voice that provide the spaces in the novel where the dynamics of metamorphosis of characters the reader knows from *Jane Eyre* can be observed. Just as the characters are seen in a new light, they are also shown to metamorphose into categories that are different from the ones they seem to belong to on the surface.

This transformation effected through intersubjectivity, I maintain, is a strategy of the mad-witch story telling mode, which I want to go into now. Kathleen Renk says that contemporary Caribbean women's writing transforms the madwoman figure we know from Victorian texts (the one that is expounded on in Gilbert and Gubar) "into a figure of rebellious, socially transformative power, a figure who does not stagnate in madness but rather uses her unconscious mind to educate and transfigure herself and others" (Renk 93). These narratives dissociate stagnation from the madwoman and point to her mode of articulation as an unofficial, alternative resource for power and agency. If the subject is denied these powers such as legitimate use of language, it will look for other, non-conventional ways, favouring modes of "revelation" (Renk 18) over linear narration, and violence over appeasement. To be able to function fully, the disenfranchised, mad connotated (in this case colonial to boot) woman needs new hermeneutic powers to make sense of experience. Indeed, Renk identifies the mad-witch in contemporary women's texts and conceptualizes it in the following way:

> The contemporary women's text often uses fantastic supernatural elements as a discursive strategy. Obeah witches with beneficent and deadly power and old, coarse storytelling hags driven by anger and vengeance are commonplace in these texts. These female figures often act as spiritual guides who remember the past while they magically transform the present. The "mad" witch storyteller perhaps best represents the iconoclastic power of this literature. Her story seems to emanate from a broader, non-transcendent vision that reshapes the solipsistic, fragmented view of the world. The use of the storyteller's method that simultaneously destroys and rebuilds is at the heart of the narrative strategies of contemporary Caribbean women's writing (Renk 8).

These characteristics are concurrent with Rhys's style as well, especially the narration of Antoinette. Antoinette resorts to obeah[1] a couple of times during the novel to get her husband back, and one can see that she tries to mimic the effects of obeah through her mesmerizing narration. This retributive narration, of telling the story from her side, is a means of getting her own back for Antoinette, a chance for her to vilify Rochester, as she herself has been vilified in *Jane Eyre*. In mad-witch narration, more often than not the restitution of speech for the "mad" or "beyond the pale" connotated figure involves retribution for the hitherto unacknowledged moments of silencing. Through her narration the "mad" character offers another epistemology where different meanings are attached to various symbols and actions. Such revision of values and at times history comes through what can be termed rhetorical violence and sometimes violence that is physically inflicted. And indeed, both in *Jane Eyre* and *Wide Sargasso Sea* Antoinette inflicts violence, by burning down the manor house. In *Jane Eyre*, this burning down is written off as an act of a madwoman, whereas in *Wide Sargasso Sea*, Antoinette lets us know the reasons lying behind this flagration as a manifestation of retribution, thus the reader is made privy and participant to an epistemology other than the one in *Jane Eyre*.

---

[1] A kind of witchcraft practiced in the Caribbean.

The metamorphosis that we are made to observe in both Antoinette and Rochester actually point to a wider historical one concerning the characteristics of the colonial categories represented by these two characters. The term "colonial" is a generic one that may refer to a wide variety of subjects who live in the "colonies." Under this umbrella term one can be referring to categories such as settler, colonial and post-colonial, however the generic one "colonial" reminds us that these distinctions are historicised. Coral Ann Howells positions the two novels *Wide Sargasso Sea* and *Voyage in the Dark* where Rhys introduces a prominently West-Indian character "as illustrating the shift from colonial to postcolonial" (Howells 23). Different identities and subjectivities develop along with historical events, conquests and emancipatory acts. In the characters of Rochester and Antoinette, I recognize the categories of settler and the creole, although other attributions could be just as legitimate. I am especially interested in the ways these categories interact and how, as they interact, they blur the boundaries. While Rochester can be categorized as the "settler," Antoinette has moved unto this differentiated identity. Martina Ghosh-Schellhorn argues that this is the "Creole," which she conceptualizes as transitional identity (178). This transitional identity, resulting from the distance felt both towards the English language and the landscape at the same time, lets us read one subject's predicament in the other. In *Wide Sargasso Sea*, the mad witch narration mode simulates the transformative space of colonial Caribbean, and enables the reader to see the transformations in these characters.

Robert Kendrick also points to Rochester's transformation and puts the stress on Rochester's particularity as a white male in the settler role and considers the novel as a space where Rochester reframes/revises his position in the patriarchal and colonial discourses:

> Though the changes that occur in the characters of Antoinette and Jane have been much noted and discussed, the efforts of feminist readings to chart Antoinette's and Jane's re-articulations of their relations to the patriarchal discourses embodied by Rochester have not explored fully the possibility that Edward himself rearticulates and redefines his position as a masculine subject, as he re-examines the ethical implications of the masculine prerogatives that he has enjoyed and abused[2] (237).

---

[2] Kendrick goes onto argue that: "This question must be addressed if the depth and potential of Bronte's ethical revision of gendered subjectivity in her "original" work, and the reaffirmation of this re-vision in Rhys's "supplement" are to be recognized, for the novels of Bronte and Rhys offer intriguing representations of how Victorian subjects lived at odds with the dominant cultural narratives of class and gender."

Indeed, it is through his remembering and narrating in *Wide Sargasso Sea* that Rochester reconsiders his position as a white male subject. He remembers his re-enactment of the unsettling experiences of the settler in the Caribbean wistfully, a remembering of which he has tried to relegate to Antoinette in the attic in the fictional world of *Jane Eyre*. For Rochester's subject, rearticulating and redefining his position is possible only through reframing his discourse concerning his marriage: that is to say a re- and fuller narration of the events that took place in the Caribbean. In *Wide Sargasso Sea* he gives the lie to his own statements in *Jane Eyre* concerning his marriage as having been a catalogue of "hideous and degrading agonies" (304), as he recollects the night when Antoinette drugs him with an *obeah* potion she has procured from Christophine, her now emancipated slave:

> I had never seen her look so gay or so beautiful. She poured wine into two glasses but I swear it was long before I drank that I longed to bury my face in her hair as I used to do. I said, "We are letting ghosts trouble us. Why shouldn't we be happy?" She said, "Christophine knows about ghosts too, but that is not what she calls them." She need not have done what she did to me. I will always swear that, she need not have done it.(Rhys 87).

Kendrick brands such passages as the "intriguing representations of how Victorian subjects lived at odds with the dominant cultural narratives of class and gender" (36). Rochester's self-confessed complicity in the sensual world of Antoinette thus can be seen as a resistance to the cultural narrative of the prude and civilized white male reigning in the wild creole. This, it seems, is one of the ways in which Rhys re-evaluates the "cruelty" of Rochester, which Rhys, in a letter, says she finds difficult to accept or understand (Angier 44). His infatuation with Antoinette sheds a different light on his marriage, transforming this relationship into one where there is a strong bond, rather than the very loose one depicted in *Jane Eyre*.

In addition to his affection for Antoinette, Rochester is also represented as being a vulnerable party according to the Victorian rules of patriarchy, in which the second son is ignored in inheritance. When he finds out about Antoinette's mother's madness he decides to write an enraged letter to his father, accusing him of knowing about the "madness" in Antoinette's family and reproaches him for sending him off to the Caribbean to marry her for her money: "I know now that you planned this because you wanted to be rid of me. You had no love at all for me. Nor had my brother. Your plan succeeded because I was young, conceited, foolish, trusting. Above all because I was young you were able to do this to me" (Rhys 104). But the following is actually what he writes: "Dear Father, Unforeseen circumstances, at least unforeseen by me, have forced me to take this decision and I am certain you will believe that the less you talk to anyone about my affairs, especially my marriage, the better. This is in your interest as well as mine (Rhys 84). Here we see the fate-determining role of the English family and how one's agency is determined vis-à-vis one's distance to it, not only in the colonies but in the "motherland" as well. This is, as Kendrick points out, a re-examination of masculine prerogatives, which includes the passage of wealth to the first born son.

In part two of the novel, Rhys accommodates both subjectivities rather than keeping them in separate chapters, drawing attention to the interaction between the two subjectivities and how they are transformed. Gillian Whitlock suggests that we need to look at intersubjective narratives in order to locate decolonizing activities in literature, or activities that lead to a deessentialized sense of the self:

> If binaries, thinking in terms of origins and authenticity, centre and periphery, and the separation into consistent and homogenous identities are fundamental to colonizing discourses, then the work of decolonization is to return ambivalence and duplicity, and to look to intersubjectivity in cultural formations and texts. All this is implied in "intimacies" (Whitlock 6).

If we want to locate the intimacies that Whitlock speaks of in *Wide Sargasso Sea*, we need to look at the points in the narrative where the narrator changes. These points enact the passage of one subjectivity to another, and thus simulate the transformation of one subjectivity into the other. The transformation becomes visible when the narration of Rochester cuts Antoinette's, and when two characters adopt certain aspects of the discourses used by the other. As Minh-ha suggests: "Whether I accept it or not, the natures of I, you, s/he, We, they, wo/man constantly overlap. They all display a necessary ambivalence [...] Despite our desperate eternal to contain and mend, categories always leak" (Minh-ha 94). Maybe the best example of this leakage in *Wide Sargasso Sea* is the first moment when Antoinette breaks Rochester's narration in the second part. It happens, as it were, unbeknownst to Rochester, while we read a chapter in a book entitled *Obeah* with him. The text trails off in three dots, and after that the first person narration resumes:

> I did not look up though I saw him at the window but rode on without thinking till I came to the rocks. People here call them Mounes Mors (the

> Dead Ones). Preston shied at them, they say horses always do. Then he stumbled badly, so I dismounted and walked along with the bridle over my arm. It was getting hot and I was tired when I reached the path to Christophine's two-roomed house, the roof shingled, not thatched. She was sitting on a box under her mango tree, smoking a clay pipe and she called out, "It's you Antoinette? Why you come up here so early?" (Rhys 67).

Christophine perceives Antoinette here almost like a changeling, and the reader is startled just as much. The fact that Antoinette is riding the horse and that she distances herself from the landscape, transferring the knowledge of the stones to the locals by saying "people here call them" (Rhys 67), accommodate a perception of this narration as still being Rochester's. This effect of one voice transforming into the other, this strategic staging of intersubjectivity is made possible through the retributive narrative force of the mad-witch narration.

As I have mentioned above, the adoption of certain aspects of the discourse of the other, for instance the way they express their feelings for the island also works towards that end. In both the remembrance of Antoinette and Rochester, the island remains alien to a great degree. Where as Rochester claims that he is the stranger on the island, Antoinette speaks explicitly about the similarity of their positions *vis-à-vis* the island, an understanding that their narratives have already been working towards:

> "I feel very much a stranger here," I said. "I feel that this place is my enemy and on your side."
>
> "You are quite mistaken," she said. "It is not for you and not for me. It has nothing to do with either of us. That is why you are afraid of it, because it is something else. I found that out long ago, when I was child. I loved it because I had nothing else to love, but it is as indifferent as this God you call on so often" (Rhys 82).

Here, the mad-witch narration that Rhys uses stages the voice of both characters to give us an understanding of the island as an independent spirit that belongs to neither of them. The other voice that could have claimed an authentic understanding of the island, that of Christophine's, gets to narrate only as embedded in Rochester's and Antoinette's narrations, which reveals Rhys's strategy to address the issue of metamorphosis, or laying bare the differences that the reader may pick up on in characters that he/she is already familiar with. It is clear that Rhys is interested in Antoinette and Rochester, and conceives of the narrative space of *Wide Sargasso Sea* as a site of transition.

Revealing both characters' certain memories which are absent in *Jane Eyre* allows for the transformation that is enacted in the way the reader perceives the characters. Thus, the archetypes of the exploitative or heroic settler and exploited or degenerate creole, get new readings and we are led to question the boundaries between these subjectivities. Rhys enlists the help of mad witch narration in order to achieve this effect, along with the strategic use of two main narrating "I", which are shown to conflate with each other. Rhys moderates their voices so as to reveal to the reader their resemblances more than their differences, to remind us that both are colonial subjects, and the way they define themselves are equally determined by the distance they feel towards "the motherland." Through various narrative tools Rhys deessentializes their voices, with particular moments where it is hard to tell which one of them is speaking. This points to a certain continuation in the white identities in the colonies, and reveals to us the metamorphosis of the white colonial subject, from colonizer to the depraved white colonial who is now herself open to exploitation like the natives of the island, or the indentured labour brought from Africa. Thus, *Wide Sargasso Sea* stages the transformations in colonial positions, and points to the ultimate porousness of categories imposed by hegemonic structures.

Works Cited

Angier, Carol . *Jean Rhys*. Harmondsworth: Penguin, 1985.

Ashcroft, Bill, Gareth Griffith, and Helen Tiffin eds. *The Empire Writes Back: Theory and Practice in Post-Colonial Literatures*. London: Routledge, 1989.

Assman, Aleida. *Erinnerungsraeume: Formen und Wandlungen de kulturellen Gedaechtnisses*. München: Beck, 1999.

Assman, Jan. *Das kulturelle Gedächtnis: Schrift, Erinnerung und politische Identität infrühen Hochkulturen*. München: Beck, 1992.

Bal, Mieke. "Introduction." *Acts of Memory: Cultural Recall in the Present*. Ed. M.Bal et al.

Hanover, NH: Dartmouth College/University Press of New England, 1999. vii-xvii.

Bhabha, H.K. *The Location of Culture*. London: Routledge, 1994.

Bronte, Charlotte. *Jane Eyre*. London: Penguin, 1994.

Ghosh-Schellhorn, Martina. "The White Creole Woman's Place in Society." *Intertextuality ad Transcultural Communication in the New Literatures in English*. Ed. Wolfgang Kloos. Amsterdam: Rodopi, 1998. 177–89.

Gilbert, Sandra, and Susan Gubar. *The Madwoman in the Attic: The Woman Writer and The Nineteenth-Century Literary Imagination*. New Haven: Yale University Press, 1979.

Howells, Coral Ann. *Jean Rhys*. Hemel Hempstead: Harvester Wheatsheaf, 1991.

Kendrick, Robert. "Edward Rochester and the Margins of Masculinity in *Jane Eyre* and *Wide Sargasso Sea*." *Papers on Language & Literature*. 30.3 (1994): 235-242.

Minh-Ha, Trin. *Woman, Native, Other: Writing Postcoloniality and Feminism*. Bloomington: Indiana University Press, 1989.

Renk, Kathleen J. *Caribbean Shadows and Victorian Ghosts: Women's Writing and Decolonization*. Charlottesville: University Press of Virginia, 1999.

Rhys, Jean. *Wide Sargasso Sea*. London: Penguin, 2000.

Whitlock, Gillian. *The Intimate Empire: Reading Women's Autobiography*. London: Cassell, 2000.

FINDING NEW SITES: TRANSFORMATIONS OF PLACE AND IDENTITY IN THE POST-APARTHEID FICTION OF NADINE GORDIMER AND ZAKES MDA

TANIA ZULLI

Abstract

The centrality of location in post-apartheid fiction is a crucial point which gives birth to interesting considerations on social and individual life. The introduction of innovative viewpoints in narrative is often rendered through a different perception of space, as well as through the presence of additional spatial dimensions intended as symbolic accounts of inner evolutions. The following paper deals with an analysis of real and imaginary spaces in two post-apartheid novels, *The Madonna of Excelsior* by Zakes Mda and *The Pickup* by Nadine Gordimer. The presence of "real" and "imaginary" sites in the two narratives fosters the function of location as a means of building one's own new personality. Therefore, the perception of space in these novels becomes a fundamental element in the construction of new post-apartheid identities.

The fostering of new perspectives in post-apartheid narrative has become a crucial point in recent years; the shift from issues concentrated on the black vs. white opposition to other attractive subject matters has given great vitality to the South African novel, since looking for new themes also means finding new epistemic values. Among the many ways employed by writers to introduce innovative viewpoints, the presentation of a different perception of space and locations – intended both geographically and symbolically – emerges as a very interesting aspect. In such a scenario, authors like Nadine Gordimer[1] and Zakes Mda[2] play a significant role. In a recent essay on post-apartheid South Africa, Gordimer has claimed the need of a modern culture able to cross boundaries. Specifically, South African narrative should avoid cultural exclusivity and open to its neighbouring realities. Writers should "seek exchanges of the creative imagination, ways of thinking and writing" and address to "a vital mixture of individuals and peoples re-creating themselves" (Gordimer 28, 2000). This necessity is essential for a renovation of literature, for long time concentrated on political happenings and racial fights. Equally, Zakes Mda has emphasized the importance of "compromise" between past traditions and present culture, and has proposed "a flexible and adaptive use of tradition that enables and facilitates life in the contemporary world" (Bell 70).

In her latest novels, Gordimer has introduced a renovation of themes, achieved through a productive confrontation with other realities. To those asking her what she would write about after the end of apartheid, she has given exhaustive narrative answers, dealing with topical themes such as criminality and violence, pollution, and the problem of AIDS. If one would like to imagine a likely answer given by Nadine Gordimer to those wanting to catch the novelty of her post-apartheid work, this could be found in Zakes Mda's words: "We are no longer writing for the struggle, we are writing for ourselves" (Donadio). After the end of apartheid, the main challenge of many South African writers seemed that of finding new, captivating themes. As a matter of fact, a more complex literary and ethical background lies behind post-apartheid fiction: the analysis of present-day society implies a close look at an evolving setting on the one hand, and a deeper interest in aesthetics and imaginative writing on the

[1] Since the institutionalization of apartheid in 1948, Gordimer has written fiction and critical essays following all the phases of its dramatic development. Her novels show a complete and detailed picture of the periods of apartheid and post-apartheid explored through the daily lives of her characters, so that history and politics are absorbed by human perception and presented to the reader. The ability of her writing to penetrate individual consciousness caught in crucial moments, as well as her commitment in the political events of her country have been worth a great number of literary awards and the Nobel Prize for Literature in 1991.

[2] Mda is one of the best known voices of South African fiction; he has written novels, plays, poems and he is involved in several social and didactic projects on AIDS. The nature of his multivalent writing is confirmed by the adaptation of his novels to theatre, film screens, jazz pieces. Mda's imaginative universe is deeply linked to the South African world of the apartheid and post-apartheid years and his writing has won him many literary prizes, such as the Commonwealth Writers' Prize and the Olive Schreiner Prize.

other. Even if authors are not constrained in protest and resistance writing anymore, the shade of the past still dominates the country and finding a new place of compromise is one of the main preoccupations of recent fiction. This site of negotiation is variously portrayed, and it involves both spatial representations and temporal dimensions.

The present work will focus on two important aspects of location in post-apartheid fiction, each of them being clearly detectable in the novels that are going to be analysed. In both novels the centrality of location leads to a rethinking of the intersection of territoriality and identity. Specifically, the alternation of real and imaginary sites serves the shaping of a new, more complex concept of space. The idealization of place also leads to its interiorization on the part of the characters; this means that any "geographical" setting is not only part of one's own conscience, but also works in changing one's own individuality. The final and main point of this paper will concentrate on the idea that place can be functional to the shaping of modern postcolonial identities. As postcolonial locations are mainly compromise lands, a consideration of personal and national identity, as well as the analysis of the complexities emerging when considering "interstitial perspectives" is necessary. As Homi Bhabha stated, "The social articulation of difference, from the minority perspective, is a complex, on-going negotiation that seeks to authorize cultural hybridities that emerge in moments of historical transformation" (Bhabha 3). In their looking for "border locations," recent South African novels are linked to the evolutions of new postcolonial individualities.

From a structural point of view, Gordimer and Mda's narratives share a similar dual construction expressed on a variety of levels. In Zakes Mda's *The Madonna of Excelsior* dualism is firstly expressed on a temporal scale, being the plot concentrated on two different periods of time, that of apartheid and that of the post-apartheid years. The novel opens with the story of the famous 1971 case in which many white men and black women were condemned for violating the Immorality Act that prohibited relationships between people of different colour. The main protagonists are Niki and her coloured daughter Popi, whose fault is that of being the clear testimony of her mother's mischievous act. By mixing words and art as narrative devices, Mda deals with racial identity and sexual encounters across the colour line, and with their repercussions on social life. The existence of the two women is caught in the narrowness of apartheid first, and in a more tranquil, post-apartheid state after; the shift to democracy is only one aspect of the extremely complex set of transitions that involves movement of symbolic locations, modes of living, identities.

Apart from temporal doubleness, the novel reveals a twofold presentation of setting which alternates continuously between a real place – the town of Excelsior – and a more ideal, abstract location, that of the paintings of the Flemish artist Father Frans Claerhout. Being more than a simple background description, the pictures represent a pivotal part of the story, as the major events of the plot are first described through paintings. The link between the two worlds is already detectable in the author's dedication:

> On 10 May 2000, together with a phalanx of my daughters, I visited Father Frans Claerhout at his studio in Tweespruit, Free State. I had always wanted to meet him. He had mentored some artists friends of mine, James Dorothy in particular. Claerhout presented me with a book on his work written by Dirk and Dominique Schwager. But first he painted a golden bird on the black flyleaf and signed his name. I dedicate this novel to the bird[3].

While underlining the connection among all arts, the dedication to the bird, that is to a painted image, introduces the juxtaposition of writing and "visual narration" as an outstanding characteristic of *The Madonna of Excelsior*. This narrative strategy is developed through the representation of two different "artistic places," that of writing and that of painting, and is supported by a plot centred on a double location which triggers off a number of symbolic interpretations. Niki poses as a naked model together with her daughter Popi for Father Frans Claerhout: it is through this act that the two women are able to meet both worlds – their own and that of the paintings – and to migrate from a ghastly real existence to the milder, happier imaginary universe painted in Claerhout's pictures. Popi is the first to come into contact with the world of the paintings, so close to her own reality and so different at the same time:

> Twenty-five years ago Popi peered from her mother's back at the white man as he warmly and masterfully daubed his broad strokes. At five she was precocious enough to wonder why the houses were all so skewed. And crowded together. She thought she could draw better houses. Her people, those she sketched on the sand in the backyard of her township home, were not distorted like her priest's. They were matchstick figures with big heads and spiky hair. But they were not distorted. Yet his very elongated people

[3] The dedication appears just before the novel's opening and it introduces the figure of the painter as more than a fictional character. Father Frans Claerhout is a Catholic priest representing the author's alter ego. Whereas Mda describes the life of Excelsior through words, Claerhout does the same thing using his paintings.

> overwhelmed her with joy. *She saw herself jumping down from her mother's back and walking into the canvas, joining the distorted people in their daily chores. They filled her with excitement in their ordinariness* (Mda 2, italics added).

Apart from the painter's world, what Popi actually sees is "her people," that is another imaginary reality created in her own mind. In order to escape the unjust life that forced her mother to hide her when she was a little girl or to burn her skin to make it appear darker, Popi feels the need to meet an ideal place where to project the hopes of a future life. As a mixed blood child, she represents the union between two realities and embodies the existence of a new "rainbow nation," as that described by Desmond Tutu, Nelson Mandela, and various members of the ANC on the point of building a new democratic country (Zulli 27).

The rainbow world portrayed in the pictures is made up of vivid licks of paint, crowded with coloured people whose skin variety does not only span from black to white but includes all the tints of a rainbow. In order to contrast the life lived in the small town of Excelsior, where skin colour still determines people's destiny, mother and daughter take refuge in the trinity's[4] paintings. It is a place of the imagination:

> Niki and Popi frolicked in the *wide-open spaces* that the trinity created for all those who loved wide-open spaces. Those who relished big skies that merged with the earth. *Eliminating horizons*. Making it impossible to determine at which point the earth ended and the sky began. It was a rapturous sight. Popi, truly coloured in red and blue patches, running among the brown sunflowers. Petals wilted and lost their yellowness. Popi naked and unevenly coloured. Not old enough to crawl. Not old enough to toddle. Yet frolicking and running in the brown field. Niki, naked and free, running after her. Popi and Niki gambolling in the field whose wilting colours formed a fading image. Like one big veronica. Until woman and infant merged with Payne's grey. And became one with it. *Disappearing into the trinity's splashes* and becoming part of the compassion they evoked. *No one would ever find them* (Mda 69, italics added).

Presented like a celebration of the new free nation expressed through colours, the painter's world is longed by all those who are looking for boundless lands. Openness, sky and earth merging, lack of definite horizons are the main characteristics of this imaginary world. It is a place of freedom, where nakedness represents the primeval condition of man. Mother and daughter end up in merging with colours and disappearing in

[4] Father Frans Claerhout is also known as "the trinity" in the novel.

them in a far away world, whose concreteness is just as remote. Place becomes an expression of conscience, a deeply interiorized state. A new identity and a new life are gained by Niki and Popi thanks to their empathy with this world of imagination. The creations of the trinity become a kind of Platonic world of forms whose main characteristic is the lack of every change or transition since everything is already perfect there. In opposition to the continuous modifications of everyday life, the universe of paintings is characterized by immobility, a state that fosters eternity.

Again, a contrast is built between the evolutions of the social and cultural life of South Africa moulded by historical events and the presence of a permanent, ideal world that stands as a point of reference, a perfect, ideal reality to be aimed at. The shift from apartheid to post-apartheid is presented through an interrelation of events and characters, to suggest that life in a country free from the laws of apartheid is at any rate influenced by years of institutionalized racial discriminations. Moreover, the new political and social scenario is characterized by a plurality of associations; after the end of apartheid, Popi gets involved in politics and becomes a member of her town council together with Viliki and Tjaart Cronje, who are both her half brothers. None of them is fully satisfied with his/her life and position; Popi in particular resents her own identity of outsider, being neither black nor white. Sometimes she feels that there is no place for her in the new post-apartheid reality, not for the moment at least, and the only refuge she can take is in the trinity's paintings that constitute a way of healing her heart (see Mda 268). The final message of the book is that there is a unifying, meaningful place where to look for the merging of all differences; this is the world of paintings, which represents, in its turn, the universe of art.

Concerned with essentially "real" locations, Nadine Gordimer's novel, *The Pickup* introduces themes that are equally linked to issues of identity and personal identification with space. The meaning of *The Pickup* is entirely contained in the few words of its opening quotation. The sentence, taken from a poem by William Plomer says: "Let us go to another country [...]/The rest is understood."[5] The whole story concentrates on the idea that life, real life, is elsewhere, and this brings to the fore the value of location in this novel and, from a wider point of view, the importance of place in postcolonial and post-apartheid fiction. Gordimer's story allows the reading of place under a double perspective:

[5] William Plomer (1903-1973) was born in South Africa but his parents were English and he lived in South Africa, Great Britain and Japan. He is author of fiction, poetry, critical essays and his masterpiece is *Turbolt Wolfe* (1926), a novel on interracial love. The poem quoted here is contained in his *Collected Poems*.

first of all as a geographical site, secondly as an element endowed with great symbolic value and therefore able to foster changes and renovations of identity. Also this novel has a dichotomic configuration, and both setting and characters reinforce its structural division. Together with the opposition of different places, Gordimer analyses the effect that moving from one place to another has on different personalities. What does transferring to another country mean? What is it like to meet another culture, different people, languages, customs? How do these encounters influence the shaping of identity? The answers to these questions can be found in the novel while reading about the lives of the two protagonists, Julie Summers and Ibrahim-ibn-Musa, who are both taken to a country which is not their own and that is completely different from their life style. Ibrahim is an illegal immigrant living in post-apartheid Johannesburg and his apparent condition of alleged hardship is not only accepted, but also strongly longed for. On the other hand, Julie's flat life in the South African city changes radically only when she has the opportunity to leave her rich, despised family to follow her partner – and subsequently husband – to his hometown. Ibrahim is forced to go back because his permit of residence has expired, while Julie's journey to the unnamed Islamic country is made to accompany Ibrahim and proves to be a highly fulfilling experience.

In presenting the journey of two characters whose lives are basically different (if not opposed), Gordimer introduces an interesting aspect of post-apartheid transitions, focusing on the manifold possibilities given by the research of an ideal place and suggesting a variety of interpretations linked to these researches. The shift from one place to the other in *The Pickup* is introduced as an escape of the two protagonists from the city of Johannesburg to a small country village near the desert. This way, the novel becomes also a narration of borders and crossings; the fact that neither of the two countries in which the story is set is clearly identified by the author hints at a generalization symbolizing a wider division between East and West, that same division recalled by Edward Said in his *Orientalism*. In Gordimer's novel though, the neat opposition East/West incurs modifications: a modern, almost Europeanized South Africa plays the part of Said's West, while the Arabic village is often described according to the stereotypes of eastern culture. Johannesburg is unequivocally a post-apartheid city of the late Nineties but instead of symbolizing a newfound ideal state of freedom, it still represents – like Mda's Excelsior – a problematic concoction of manifold realities.

The journey to the small village near the desert is the metaphor of a voyage into the characters' inner self, a discovery of their own real individuality. Every kind of relationship is felt by the two protagonists in terms of place recovery or possession. Also the field of sexual interaction is the "place" to look for happiness and reciprocal understanding, and even love is a utopian nation where cultural totality can be achieved. However, Julie and Abdu cannot find an ideal common dimension existing outside the sphere of sexuality; they both tend to another country, but it is not the same country they want to share. They are aware that "there is no future without an identity to claim it" (Gordimer 37, 2001), and they are moved by a sense of lack they feel inside themselves, but each of them finds his/her ideal place in the country of the other, so that they seem to move in opposite directions. This happens because they start from opposite realities and tend towards opposite targets. They move together only physically, while their inner wishes track divergent directions.

While Abdu spends his days trying to obtain a permanent residence in any of the so-called civilized western countries, Julie feels like she is living a sterile life in the city; it is a reality she does not belong to, and she continuously delineates herself in opposition to her culture. Once in her husband's village, she rises towards her self-realization and gradually builds her new identity willingly accepting the rules of a different life.

As it happened in *The Madonna of Excelsior*, the importance of place in *The Pickup* is mainly linked to its interiorization on the part of the characters. Sites and locations acquire significance according to the role they play on individuality. Julie reaches a state of communion with the surrounding natural landscape when she takes solitary morning walks in the desert near the village. The desert holds the same characteristic of materiality and idealism contained in Father Frans Claerhout's paintings. It is a place whose sights and silence plunge deeply into conscience and are easily interiorized:

> The desert. No seasons of bloom and decay. Just the endless turn of night and day. *Out of time*: and she is gazing – not over it, taken into it, for *it has no measure of space, features that mark distance from here to there*. In a film of haze there is *no horizon*, the pallor of sand, pink-traced, lilac-luminous with its own colour of faint light, has *no demarcation* from land to hair. Sky-haze is indistinguishable from sand-haze. All drifts together, and there is no onlooker; *the desert is eternity* (Gordimer 172, 2001, italics added).

Just as in Father Frans Claerhout's picture, openness, undefinedness, lack of horizons convey the image of an ideal world that is only partially felt through the senses, its real essence contained in the heroine's mind. The lack of boundaries makes it a place suitable for mental assimilation; its stillness makes it, again, the Platonic world of forms where nothing is subject to evolutions, being things already perfect in their state. In this half-real, half-imaginary site Julie undergoes a personal metamorphosis, an identity change deriving from the interiorization of place that hints at a spiritual dimension. As a consequence, the desert becomes the place where a new, more stable identity is born. It represents an out of time condition characterized by perpetuity and opposed to the transitoriness of everyday life.

Julie's choice to embrace this world and live in her husband's village even after he goes to the U.S. is the metaphor of a regained sense of individuality that she obtains after going through a rite of passage typical of Gordimerian heroines. When in Abdu's village, Julie feels like living "in a common existence-that-does-not-exist [and] she has no sense of who she is in this immersion, everyone nameless […]" (Gordimer 168, 2001). When in this new, foreign place, the woman seems to have cancelled all the signs of her previous identity and starts building herself as "a new person." There seems to be a kind of reciprocal influence between the woman's sensibility and the place she finds herself in; if location stimulates a new attitude in line with Julie's wishes, at the same time, landscapes are dependent on the personal interpretation she gives of them. If new sites exist for people as a "projection of human sensibility upon the natural or built environment" (Kennedy 5), it means that there is no place apart from those interpreted by our consciousness. The perception of this new space is so close to Julie's mind that it fosters her integration, which she tries to achieve by means of will and commitment. Julie adapts to her new way of living reading the Koran and learning the language because she understands that religion and speech are the most important elements for her socialization. In an age of migrations, displacement and change, language symbolizes contact, communication, integration. Linguistic instruments are the main way to take part to a reality Julie perceives as extraordinarily rich in values. Despite everyday life limitations, she sees the village as her new, long-awaited home, where she can achieve her own inner spiritual harmony. The places surrounding her are both real and imaginary; at any rate in perfect agreement with her own individuality. They are "home" in the sense Gaston Bachelard defined it: "All really inhabited space bears the essence of the notion of home […] the sheltered being gives perceptible limits to his shelter. He experiences the house in its reality and in its virtuality, by means of thought and dreams" (Bachelard 5).

The brief analysis of these two possibilities of framing the concept of location in new South African fiction brings to the fore important considerations on the idea of space and its transformations. First of all, a distinction is made between a down-to-earth notion of place as an evolving, complex reality and an ideal, more abstract kind of location which is mainly constructed through aspirations, desires, and hopes. This second case reflects the spatial dimension writers introduce as a new possibility in post-apartheid literature. Moreover, these more ideal sites are strictly linked to conscience and often intended as projections of the mind. This leads to the necessity felt in recent fiction of building a "better place" that can be reached by people only by taking into account all the implications deriving from a complex world.

The formation of a new postcolonial society as that described by Homi Bhabha in *The Location of Culture* takes the move from a reconsideration of space and locations intended as evolving entities, subject to the influence of time, political events, individual changes:

> [...] in the *fin de siècle*, we find ourselves in the moment of transit where space and time cross to produce complex figures of difference and identity, past and present, inside and outside, inclusion and exclusion. For there is a sense of disorientation, a disturbance of direction, in the 'beyond': an exploratory, restless movement caught so well in the French rendition of the words au-delà – here and there, on all sides, fort/da, hither and thither, back and forth.
>
> The move away from the singularities of 'class' or 'gender' as primary conceptual and organizational categories, has resulted in an awareness of the subject positions – of race, gender, generation, institutional location, geopolitical locale, sexual orientation – that inhabit any claim to identity in the modern world. What is theoretically innovative, and politically crucial, is the need to think beyond narratives of originary and initial subjectivities and to focus on those moments or processes that are produced in the articulation of cultural differences. These 'in-between' spaces provide the terrain for elaborating strategies of selfhood – singular or communal – that initiate new signs of identity, and innovative sites of collaboration, and contestation, in the act of defining the idea of society itself (Bhabha 2).

Novelty comes out of complexity: it is through the elaboration of cultural difference that new social and political categories are created. The forms of space presented in post-apartheid fiction profit from this articulation and become in their turn more and more composite and multifarious. On the other hand, the shaping of new identities built on a ground of compromise and cooperation benefit from the construction of "innovative sites," be they either ideal places of the self or concrete geographical locations. The reciprocal influence of space perception on human life and of individuality on places gives birth to original structures that are presented as concrete locations, imaginary lands, or inner perceptions.

Works Cited

Bachelard, Gaston. *The Poetics of Space*. Boston: Beacon Press, 1958.

Bell, David. “The Persistent Presence of the Past in Contemporary Writing in South Africa.” *Current Writing* 15.1 (April 2003): 63-73.

Bhabha, Homi. *The Location of Culture*. London: Routledge, 1994.

Donadio, Rachel. “Post-Apartheid Fiction.” *The New York Times*. 3 Dec 2006 downloaded on April 8th 2007. <http://www.nytimes.com/2006/12/03/magazine/03novelists.html?pagewanted=1&_r=1>.

Gordimer, Nadine. *Living in Hope and History: Notes from Our Century*. London: Bloomsbury, 2000.

—. *The Pickup*. London: Bloomsbury, 2001.

Kennedy, J. Gerald. *Imagining Paris. Exile, Writing, and American Identity*. New Haven: Yale UP, 1993.

Mda, Zakes. *The Madonna of Excelsior*. Oxford: Oxford UP, 2002.

Plomer, William. *Collected Poems*. London: Jonathan Cape, 1973.

Zulli, Tania. *Nadine Gordimer. Strategie narrative di una transizione politica*. Napoli: Liguori, 2005.

ARMED, DANGEROUS, AND WALKING THE STREETS: THE WOMEN OF FANON'S "ALGERIA UNVEILED"

MAZI ALLEN

Algeria (Un)veiled

At the beginning of *A Dying Colonialism*, Frantz Fanon notes the significance of clothing in defining civilizations:

> The way people clothe themselves, together with the traditions of dress and finery that custom implies, constitutes the most distinctive form of a society's uniqueness, that is to say the one that is the most immediately perceptible. Within the general pattern of a given costume, there are of course modifications of detail, innovations which in highly developed societies are the mark of fashion. But the effect as a whole remains homogeneous, and great areas of civilization, immense cultural regions, can be grouped on the basis of original, specific techniques of men's and women's dress (Fanon 35, 1965).

The wearing of the veil (or *haïk*, as it was called in North Africa) was an example of such a "tradition" (ibid 36n4). However, as such, it came not only to identify the society in which it existed but also to mark its wearer, in particular, as an "Arab Woman" (ibid 36). This, in turn, had taken on a negative significance since the occupation of Algeria in 1830 – signifying the "Moslem woman" as "humiliated, sequestered, cloistered [...]" (ibid 38). The "Arab" or "Moslem" woman, as "she who hides behind the veil," was seen as completely, totally, lacking in political agency (Fanon 36, 1965). Thus, using the supposed oppression of women by Islam as pretext, the French colonizers of Algeria began a campaign of "liberating" the "Moslem" woman from the "veil" (ibid 38-39). However, even in everyday life, perceptions of the Algerian woman's lack of agency were not uncommon.

For instance, according to Fanon, it was quite common for the Algerian man to be questioned by European co-workers about his wife "wearing the veil": "Does your wife wear the veil? Why don't you take your wife to the movies, to the fights, to the café?" (ibid 39). Or better, his employer would put the question differently: "The firm being one big family, it would be unseemly for some to come without their wives, you understand [...]" (ibid 40). And of course the Algerian man "understood" all too well – such "invitation" was meant to put him in the untenable position of either conforming to the colonizers' expectations and stating "My wife wears the veil, she shall not go out" or else of "prostituting his wife" to his employer's voyeurism "Since you want to see her, here she is[!]" (ibid 40). In either case, the effect was to render the Algerian woman an object exchanged by men (reminiscent of Irigaray's later account).[1] Given this perception of the Algerian woman, it is quite understandable that (upon seeing veiled women in Algiers) a Frenchman would state that the Algerians were "concealing so many strange beauties" and that, having a "cache of such prizes, of such perfections of nature," they "owed it" to themselves "to show them, to exhibit them" – even stating "it ought to be possible to force them to do so" (ibid 43). Noting this, Fanon observes:

> [...] Jean-Paul Sartre, in his *Réflections Sur la Question Juive*, has shown that on the level of the unconscious, the Jewish woman almost always has an aura of rape about her.
>
> The history of the French conquest of Algeria, including the overrunning of villages by troops, the confiscation of property and the raping of women, the pillaging of the country, has contributed to the birth and crystallization of the same dynamic image. At the level of the psychological strata of the occupier, the evocation of this freedom given to the sadism of the conqueror, to his eroticism, creates ... fertile gaps through which dreamlike forms of behavior and ... criminal acts can emerge.
>
> Thus the rape of the Algerian woman in the dream of the European is always preceded by a rending of the veil. We here witness a double deflowering. Likewise, the woman's conduct [in these dreams] is never one of consent or acceptance, but of abject humility (ibid 45).

Referring to examples conveyed to him in his clinical capacity, Fanon also found the dreams of these colonizers quite revealing of the social

---

[1] I refer to Luce Irigaray's "Women on the Market" in *The Sex Which is Not One* (Cornell UP, 1985). However, other treatments such as María Lugones's "Heterosexualism and the Colonial/Modern Gender System."*Hypatia* 22.1 (2007), might be more relevant in the case of Algerian women under French colonialism.

milieu of colonial Algeria. Whether surrounded by a "harem" of submissive "Arab" women or rending the veil of an Algerian woman while forcing himself on her, the dreams of such colonizers revealed an obsessive, even sociopathic, desire to control, dominate, and oppress (ibid 46). This in turn revealed, for Fanon, the falsehood of the supposed "concern" of the colonizer about "the fate that [the Algerian] metes out to women" (ibid 38). For him, this "concern" was yet another example of the colonizers' psychological "projection" of their desires onto the colonized – a phenomenon he had described in a previous work (ibid 165): the colonizer desired to make the Algerian woman the passive, dominated, subject whom he outwardly sought to liberate. "Every veil that fell, every body that became liberated from the traditional embrace of the *haïk*, every face that offered itself to the bold and impatient gaze of the occupier, was a negative expression of the fact that Algeria was beginning to deny herself and was accepting the rape of the colonizer" (ibid 42).

Thus the Algerian women already had more than a few reasons to be interested in decolonization. Hence it is not surprising that when the opportunity presented itself – with the commencement of hostilities in November of 1954 (ibid 65) – these supposedly oppressed, repressed, meek, mild, submissive "Moslem" women became some of its most active, creative, even inventive, participants.

"A Wholly Revolutionary Step ..."

Disguised as a fully assimilated "unveiled" Algerian woman, the Algerian woman who participated in the decolonization struggle deftly slipped by the French authorities who would only see her as one of those willing to "deny herself and accept the rape of the colonizer" (ibid 42). Thinking that she was entering the European section of Algiers merely as a prostitute or other woman of loose morals, the guards at checkpoints between the segregated "Arab" and "European" sections of Algiers would not suspect in the least that she was carrying "three grenades in her handbag" or a report on police activity for the area in her bra (ibid 50). Feeling "improperly dressed, even [...] naked" (ibid 59) in the fashionable attire of 1950s France, the newly "unveiled" resistance fighter had to calmly maneuver through colonial Algiers "like a fish in the Western waters" (ibid 58) – she had to take on the role of the "amoral and shameless" unveiled woman (ibid 44n8) while, at the same time, not drawing too much attention to herself (ibid 59).

In accomplishing her mission, then, she would employ the space associated with her body to convey "complex symbolisms, sometimes coded, sometimes not, linked to the clandestine or underground side of social life" (Lefebvre 33). This level of space – which Henri Lefebvre would later term "representational space" – was "directly *lived*" by the Algerian resistance fighter through her body, "its associated images and symbols" at the same time overlaying it and giving it significance (ibid 39). As Fanon describes how she negotiated the level of her own bodily space (Fanon 58-59, 61-62, 1965), he notes that she had to also negotiate the space of the modern "European" section of Algiers which literally confined, isolated, "surrounded … [and] laid siege to" the segregated "Arab" section (ibid 51-52).

Moving through Algiers, she would have to maintain extraordinary self-control – for "walking the streets" she would encounter the same objectification that a "real" assimilated Algerian woman would: "An unveiled Algerian girl who 'walks the street' is very often noticed by young men who behave like young men all over the world, but who use a special approach as the result of idea people habitually have of one who has discarded the veil" (Fanon 53, 1965). This "special approach," again, was the one used for prostitutes or other women of supposedly loose morals. However, as she was taking on such an abject persona *as* her own, the resistance fighter would have to walk on – even if gritting her teeth at the "unpleasant, obscene, humiliating remarks" (ibid 53) which would surely follow – all the while making sure the articles she carried arrived at their destination.

However, inevitably, rumors would filter back to her male relatives that "Zohra or Fatima" was seen in the downtown section of Algiers, unveiled and "walking" the streets. The "dishonored" father (or brother or spouse) would confront her on this unseemly behavior, only to find that she was indeed walking the streets–but not for the reasons he had imagined (ibid 60). Thus, after considering her profound commitment to the liberation struggle–even in the face of the certainty that "any Algerian woman arrested would be tortured to death" (ibid. 49)–he would think better of his initial outrage and give his formal consent to her activities.

Fanon thus goes on to describe even of how, for purely tactical reasons, the veil was "Removed and reassumed again and again" (ibid 61). Again and again, the women readapted their body images towards the stereotypical views of the colonizer: this time as the "submissive" housekeepers "indiscriminately" called "Fatma" (ibid 52) the next as the "shameless" assimilée. Taking advantage of the weaknesses inherent in the colonizer's efforts at profiling, the Algerian women ended up destabilizing the colonial regime in ways that were previously unimaginable: "All historic limits crumbled and disappeared" (ibid. 61).

Concluding his account, Fanon notes how experts in the social sciences had justified continued colonization, and especially the campaign to "liberate" women from their clothing, by demonstrating – with empirical data, field studies, and such – how "oppressed" the "Moslem" woman was (Fanon 64, 1965). To such contentions, Fanon opposed the actions of the women-as-resistance-fighters – whom he had also observed firsthand. Of them, he states "In forty-eight hours the Algerian woman has knocked down all the pseudo-truths that years of 'field studies' were believed to have amply confirmed" (ibid 65). Further:

> [...] there is the much discussed status of the Algerian woman – her alleged confinement, her lack of importance, her humility, her silent existence bordering on quasi-absence. And "Moslem society" has [supposedly] made no place for her, amputating her personality, allowing neither development nor maturity, maintaining her in a perpetual infantilism.
>
> Such affirmations, illuminated by "scientific works," are today receiving the only valid challenge: the experience of revolution (ibid. 65-66)

The Limitations Imposed by History

Although it is quite remarkable that Frantz Fanon, of all people, would take the "wholly revolutionary step" (ibid. 49) of even noticing the activities of women (let alone attribute any importance to them), his rather forward-thinking account of their activities still contained blind spots. Given the perspective from which Fanon was writing his works, such blind spots (and the conflicts arising from such) are indicative of productive thought. However, all too often they are ignored by subsequent readers seeking to deify the great thinker and insisting that any error or conflict detected in his thinking must be the result of bad translations. Yet, if we are to pay homage to thinkers like Fanon, we should do so by pointing out their inconsistencies and the limitations their social milieu imposed upon them. Hence we return to the question of agency, for here we find significant examples of Fanon's blind spots.

For instance, in speaking of their formation as resistance fighters, Fanon attributes to the knowledge of urban geography had by these women, as well as their inventive strategies of maneuvering around the colonizer, to mere "instinct" (50). Further, he even insists that "even [at the age of] twenty or twenty three" the Algerian woman "hardly ever has the occasion to leave the family domicile unaccompanied" (49). Although the aforementioned views point directly to a history of resistance in

Algeria–where, as late as 1847 (seventeen years after the French conquest), women actually did "refuse to work as domestic servants" (Lazreg 45)–and to the desire to portray Algerian women as virtuous (thus never seen unaccompanied in the European section of any city in French Algeria), both accounts were simply implausible. Fanon himself knew this well.

The women Fanon described, in fact, did venture outside the "family domicile" in search of work – as housekeepers, maids, and other service workers. This was even the case for women well under the age of "twenty or twenty-three." As Fanon himself acknowledged (in an earlier work), girls as young as fourteen were routinely subject to sexual harassment while working as housekeepers (Fanon 46n5, 1967). However, Fanon's unease with his own assertions to the contrary become quite apparent when we notice that he later terms the "confinement" of Muslim women "alleged." Therefore, his assertion that the knowledge these women had of the colonial urban geography was "an authentic birth in a pure state, unaccompanied by instruction" (Fanon 50, 1965) must also be viewed as questionable. Their own experiences working in the heart of the colonial city provided Algerian women with more than enough training in the psychology of the colonizer–training that could scarcely be obtained elsewhere.

Thus, although it cannot be denied that Fanon was committed to objectively recording his own observations–especially in light of his medical training–just as apparent are his own sociohistorical commitments. As a male, middle-class, professional living during the 1950s–entirely French in upbringing–there were things that, despite his commitments to the contrary, he could not possibly feel at ease in describing. Seeing women as practicing agency outside of the prevailing societal norms was most probably one of those things. However, this contradiction between Fanon's objective stance and his sociohistoical commitments was precisely the thing that made "Algeria Unveiled" possible. Fanon was consciously attempting to resolve the tension between his own perceptions of the woman's role in the decolonization struggle and the rather atypical activities of certain women plainly visible to him. To ignore this aspect of "Algeria Unveiled" would render "Algeria Unveiled" an inexplicably anomalous, even redundant, piece in a work which (by the standards of the day) had already said enough about women in its section on "The Algerian Family."

At this point, then, it will be helpful to view things from another perspective in order to reveal what precisely it was that Fanon may have missed in his account of the women in the Algerian War. Thus, in light of newer accounts of agency produced since his death, we will want to re-examine “Algeria Unveiled.” One such account is found in a recent article by feminist philosopher María Lugones entitled “Tactical Strategies of the Streetwalker.”

Streetwalker Theory

Seeking “to complicate de Certeau’s description by [...] moving a disruption of the tactic/strategy dichotomy” (214), Lugones notes that de Certeau’s account of strategy versus tactics tends to render truly resistant acts (that is, acts performed outside of official channels) invisible. Although subjects of oppression are by no means “passive” within de Certeau’s view (213)–being able use systems of domination “they had no choice but to accept” for “ends and references foreign to” such systems (de Certeau xiii)–the ability to strategize, to think things through, to “theorize” were the preserve of “the powerful [...] those who act as managers, foremen, lesser officials, and upholders of [the] institutional ‘apparatus’” (210). At best, those who resisted could only hope to “insinuate” themselves “into the other’s place, fragmentarily, without taking over [the institutional apparatus] in its entirety ...” (214). To complicate and address the limitations of this view of resistance, Lugones introduces the idea of the “streetwalker theorist”–a resisting agent who plans, strategizes, thinks from “the concreteness of [...] engagement” (207). This would entail several things.

First the notion of individual–or even monolithically collective–agency would have to be abandoned in favor of a view of agency as a “practice of sustained intersubjective attention,” as “looking for company” (221-22). This is because the previous, individualistic, view obscures the fact of the institutional “backing” which sustains all supposed instances of “single authorship, individual responsibility, individual accountability, and self-determination” (210). Although individualistic “agency” leads one to believe that “if one intends into the hegemonic system’s vein, one is the author of those intentions” (217), this is exposed as a fiction as soon as one attempts to exercise agency in a different vein. “Agency” then is usually “disallowed as moral” (211) or seen as “criminal, deviant, abnormal, in need of containment, or alternatively, as frivolous and inconsequential” when viewed from “realist” perspectives (215, 230).

Hence, the agency Lugones proposes is also a form of agency which takes the social fully into account. In this view, we “speak of the ‘I’ [...] as moving through multiple encounters [and] geographico-historical paths” (226), of an “I” in which intentionality continuously transitions and disperses into the "we" (227-29). Such agency would be one which openly "understands the strategy" implicit within "the tactic/strategy dichotomy" (225). Therefore she says of such an agent (whom she designates as the *callejera* or "streetwalker"):

> La *callejera* looking for company focuses most intently on the hostilities, discontinuities, and possibilities of connection and sense making among *las/los atravezadas/os* – "those who cross over, pass over, or go through the confines of the 'normal.'"
>
> The walking/theorizing [of la *callejera*] defamiliarizes "common sense," as the "common" backing it up loses its unseen, or taken for granted, part of the furniture of the universe, quality [...].
>
> La *callejera* can both sense the reality and fiction of the common in daily interaction as well as the multiple contestations of the common and of its sense. (221-22)

As she "hangs out" in various worlds of sense, consistently, persistently, crossing the "bounded territories that mythologize sameness," the streetwalker at the same time cultivates her ability to "create" and thrive in these different social terrains in the company of others (220-221, 224). "The streetwalker theorist," as one who exercises agency, thus "walks in illegitimate refusal to legitimate oppressive arrangements and logics" (221). As she moves through various worlds of sense, she is constantly asking herself "Within which conceptual, axiological, institutional, material set of limitations is the meaning of the possible being construed?" (231). Although this very short summary of Lugones's article certainly does not do justice to it, the conception of agency described seems to be quite relevant to Fanon's discussion in "Algeria Unveiled."

Armed, dangerous, and on the loose

For instance, in reading Fanon's account of Algerian women during the decolonization struggle, one is struck by how they indeed inhabited and created various worlds of sense. Knowing fully what was expected of her from "oppressive arrangements and logics," she tactically "insinuated" herself into the roles laid out for her – openly and consciously aware of their fictitious quality. Whether as the submissive housekeeper or that "shameless," unveiled, so-and-so-walking-the-streets, she played her role to perfection – adjusting the representational space of her own body-image according to either the latest fashions out of Paris or the most traditional of outer garments.

In an era where the dominant ideal of the women was the housewife (even in the modern West), these women openly, flagrantly, flaunted "common sense" (221). While the revolutionaries of the Front de Libération nationale[2] only hesitantly sanctioned such activities, the women themselves (certainly "in company" with one another) devised strategies of their own, contributing from their various walks of life and experiences to a strategizing of resistance from the concreteness of their own engagement with the struggle. In fact, it was the initial commitment of the women to the Revolution (Fanon 51, 1965) which eventually led the FLN to its "revolutionary step" of involving "women as active elements of the Algerian Revolution" (ibid 48-49). For this reason, in documenting the five years of the Algerian struggle (up to 1959), it should be no surprise that Fanon found their strategic "innovations [...] of particular interest" (ibid 47). These women Fanon described certainly did fit Lugones's description of resistant agents in that they challenged the "conceptual, axiological, institutional, material set of limitations [...] of the possible" (231) imposed upon them at all levels.

Works Cited

Certeau, Michel de. *The Practice of Everyday Life*. Steven Rendall, trans. Berkeley: University of California Press, 1988 [1984.]

Fanon, Frantz. *A Dying Colonialism*. Haakon Chevalier, trans. New York: Grove, 1965 [1959].

—. *Black Skin, White Masks*. Charles Lam Markmann, trans. New York: Grove, 1967 [1952].

Irigaray, Luce. *The Sex Which is Not One*. Catherine Porter, trans. Ithaca: Cornell UP, 1985.

Lazreg, Marnia. The Eloquence of Silence: Algerian Women in Question. New York: Routledge, 1994.

Lefebvre, Henri. *The Production of Space*. Donald Nicholson Smith, trans. Oxford: Blackwell, 1991 [1974].

Lugones, María. *Pilgrimages/Peregrinajes: Theorizing Coalition Against Multiple Oppressions*. Lanham, MD: Rowman & Littlefield, 2003.

[2] Commonly known by its acronym "FLN," this was the primary independence movement during the Algerian War and later Algeria's first post-independence political party.

THE DESERT AS CATALYSIS: A READING IN THE PERSONA OF AL-SHANFARA'S "LAMIYYATU'L ARAB"

VISAM MANSUR

Abstract

The desert as a place, like the sunrays that grill the granules of desert's sand, is hardly affected or metamorphosed by those who travel through it. It is the traveler who is more often transformed under the impact of the place. al-Shanfara, the author and the persona of "Lamiyyatu'l Arab" (a pre-Islamic 68-line, sixth century Arabic poem), in an attempt to leave his tribe, undertakes a journey to an unknown destination. The major bulk of the highly eloquent and archaic verses describe the poet struggling with the place he is moving through. Though the poet appears to consciously deny any identity change as a result of his interaction with the place, in the process of his journey the place radically mediates his sulphuric personality.

Introduction

The desert as a place, like the sunrays that grill the granules of the desert's sand, is hardly affected or metamorphosed by those who travel through it. It is the dweller or the traveler who is more often transformed under the impact of the place. Al-Shanfara[1], the author and the persona of

---

[1] Little is known about al-Shanfara of Azd beyond what his surviving poems tell. It is, however, certain that the poet belonged to a stock of the Yemenite tribe called al-Azd, and probably to the clan of it called al-Hijr. His father's name is not mentioned, although according to a verse attributed to him in *al-Aghani* by al-Isbahani, he claimed to be of noble parentage on the side of both father and mother. This seems unlikely; probably he had African blood in his veins, since one of the interpretations given of his name is that it means 'thick-lipped'. On the

"Lamiyyatu'l Arab"[2] in an attempt to leave his tribe, undertakes a journey to an unknown destination. The major bulk of the highly eloquent and archaic verses describe the poet struggling with the place he is moving through. Though the poet appears to consciously deny any identity change as a result of his interaction with the place, in the process of his journey the place radically mediates his phlegmatic personality. The paper will highlight the effect of interfacing with the desert on the identity of al-Shanfara and examine the dialectical interconnectivity between place and identity in the light of al-Shanfara's experience.

Al-Shanfara initiates his poem by taking a decision to abandon his tribe for reasons of his own. Among the reasons that lurk behind his strong desire to abandon his kinsmen so urgently are a deep sense that he is not appreciated among his people, his extreme sense of superiority coupled with a deep conviction that his people are inferior to himself, and an extreme sense of individuality that ridicules his current social mores. As a matter of fact Al-Shanfara's account of his interaction with the desert goes through five phases:

---

whole, al-Shanfara belongs to a group of poets known as the brigands or Sa'alik. These poets thrived on defiance, alienation, and extreme contempt for tribal mores. Their poetry, assert Beeston et al., tells of their wild life as they sang defiantly of their "high ideals of liberty and exemplary conduct" (33). Ilse Lichtenstadter (24), though she sees in these poets a form of rising controversial individuality as opposed to mainstream subjectivity of the aggregate society, implicitly hints that al-Shanfara's individuality is tarnished by the fact that he adopts the conventional form and the regular style of mainstream tribal poets. For more details about al-Shanfara's life See Charles James Lyall, 2:68-69; al-Isbahani, xxi, 134.

[2] A pre-Islamic 68-line, sixth century Arabic poem. For the original text of the poem in Arabic and its classical explanation, See Muhammad ibn 'Umar al-Zamakhshari (40-78). As a matter of fact, many scholars have disputed the authorship of the poem. Taha Hussein, among others, expresses his suspicion of the authenticity of most of the so-called Jahiliyyah poetry. See Taha Hussein. Stetkevych points out various sources that contested the authorship of the poem which on the main attribute it to the Basran poet-transmitter Khalf Al-Ahmar (d. 180/796). See Suzanne Pinckney Stetkevych (119-23). Regardless of who confected the poem, what matters most for the purpose of this paper is the text itself, as an organic entity that reflects the moods and habits of its inhabitants. Al-Shanfara, however, is the authentic persona of "Lamiyyatu'l Arab" and it is his persona as painted vividly in this poem that I intend to discuss in the light of his interaction with the place.

Interactions with the Desert

In the first three lines, Al-Shanfara expresses his heart felt desire to leave his tribe in the middle of the night to unknown destination. In the fourth line, the poet swears by the lives of his readers: "By your life, no harm on earth will befall a man who travels by night, /Whether by free will or force, if he has his wits about him."[3] He then, asserts his willful decision to leave his tribe and kinsmen for another family. His new kinsmen, as he asserts in line 5 are " a wolf, swift and sleek, a smooth and spotted leopard and a long-maned hyena." Al-Shanfara then highlights his positive personal traits in an attempt to show his readers that he is of a much noble caliber than his new kinsmen. For instance, in spite of the ferocity of the wild beasts he opts to keep company with, he tells us that he is superior to them in hunting and chasing the hunt; and when it comes to consuming the hunt, he claims "not to be the quickest" (L 8) in helping himself to the food as a sign of his graciousness. Then he devotes ten lines from 11-20 to develop his concept of masculinity, as he perceives it in himself: strong body, brave heart, dexterity at using arms, generous, and macho. This section of the poem is concluded with his hyperbolic assertion that "when the ground, hard and flint-strewn, strikes [his] hoofs, sparks and splinters fly." (L. 20) It transpires that in the first phase (L. 1-20) the poet expresses his unequivocal desire to move away from his tribe, as he is convinced of his superiority to his kinsmen whom he regards as decadents, hypocrites, cowards and effeminates.

Unlike social human beings who move in the company of others, al-Shanfara, full of self-confidence, moves alone. Ironically, he unambiguously communicates his intention to substitute his tribe with another tribe of beasts. His arrogance is not only directed at his kinsmen as he also maintains a superior attitude to his wild hosts and companions. Naturally, his expressions of superiority make sense, at the beginning of his journey, as Al-Shanfara is still endowed with health and the material comfort he took along with him upon departure.[4]

---

[3] For the English translation of this and the subsequent verses of "Lamiyyatu'l Arab" see Suzanne Pinckney Stetkevych (143-150).

[4] Due to the ambiguity in line 2 one could conclude that provisions have been prepared for his journey, too. Stetkevych in her translation of the line resolved the ambiguity by assigning the preparation of provision for the departing tribes men and not for al-Shanfara. Her English line reads: "The provisions have been readied, the night is moonlit, /The mounts strapped and saddled for your journeys ends." (142) In fact, the Arabic text does not have the possessive "your." It simply says that provisions have been readied on mounts for ends. As a result, the ambiguity of

Hungry in the Desert

In this phase (L21-41) Al-Shanfara's supplies of food hinted at in the second line – "The provisions have been readied"- are exhausted. Thus the poet enters a stage of prolonged hunger and tries his best to ignore the pain of starvation. In his attempt to do so, he swallows soil to appease his agonized entrails. Eventually, the poet is reduced to the status of a starving, wobbly wolf, hungry and thirsty that races with a formation of sand grouse landing on a spot of water to drink first. Though the poem reveals the gradual deterioration in the Al-Shanfara's social, psychological and physical status; and that life in the desert is so hard and self-dilapidating for him; Al-Shanfara still arrogantly maintains his claim to high moral grounds, honor, authenticity, altruism and chivalry.

Pride and Suffering

In this phase (L. 42-53) Al-Shanfara reveals with gruesome precision the weakness of his body as a result of the severe hunger and hardships he has been enduring in the desert. His body is transformed into a heap of fleshless bones:

> I am familiar with the earth's face
> When I take it as my bed
> On a firm back raised
> By desiccated vertebrae
> For a cushion I take an arm, fleshless,
> Its joints like gambler's bones
> When he casts them forth,
> So they stand out (L. 42-3).

Due to his weakness, he sees himself as a source of shame and frustration to war as he implicitly recalls his previous glory in the battlefield:

> Then if War, the Mother of Dust,
> Is grieved at al-Shanfara,
> Yet her delight in him before
> Was longer (L. 44).

---

the expression paves the way for another reading that assumes provisions are prepared for him, too.

Furthermore, his loneliness in the desert conjures up all the demons of the past that arrive like a swarm of deadly bees to haunt his dreams and torment his soul:

> I am an outcast hunted by crimes
> That draw lots for his flesh;
> The winner gets the first choice
> From his carcass. (L. 45)

Also, his prolonged passage in the desert with no food, drink or shelter reduced his status to that of a snake slithering naked on the burning desert sands:

> You see me, like the snake,
> The sands daughter,
> Exposed to the sun, weakened,
> Barefoot and shoeless, (L. 49)

In spite of all these evils, Al-Shanfara, refusing to admit defeat, puts on a show of bravado by convincing himself that he is still in possession of many honorable traits such as patience, content, wisdom and self-esteem that his fellow tribesmen do not have:

> Know that I am the master of endurance;
> I don its cloth
> Like a shirt on a heart like a young wolf's;
> I am shod with determination (L. 50).

Metamorphosis into a Savage

In this phase of his passage into the desert (L. 54-66) Al-Shanfara colorfully illustrates his metamorphosis into a full savage in the wilderness. After the sad experience of burning his beloved well-crafted and decorated bow and arrows to heat his weak and embattled body, he turns into a deadly beast driven by sheer instinct for survival. He becomes one of the desert marauders whom Charles M. Doughty terribly labels as "the demons of this wild waste earth, ever ready to assail passengers."[5] He attacks a caravan at night in a demonic way and "widows women and orphans children" (L. 57) and thus loses all claims to any moral ground he has talked off earlier in the text. Then in tatters, under the burning heat of

---

[5] P. 33

the sun and the gushing winds of the desert, in deplorable bodily condition, and extreme dirtiness he traverses a desert from one end to the other evading and dodging his illusionary or /and real pursuers.

Many a day of the Dog Star,
When the heat waves melt,
And the vipers writhe restlessly
On the scorching earth,
I faced straight on,
No covering to shield me
Nor any veil, except
A tattered cloak,
And a full long hair;
When the wind blows
Its matted dreadlocks fly up on all sides,
Uncombed.
Long since the touch
Of balm or delousing,
It is full and caked with filth;
For a full year unwashed.
Many a windswept plain, like the back of a shield,
A barren waste,
Its back not to be crossed,
On my two legs I crossed.
I joined its end to its beginning,
Looking out from a summit,
Now sitting, knees drawn up;
Now standing. (L. 61-66)

The Final Metamorphosis

In the fifth and last phase of the journey (L. 67-68) the last two lines abruptly terminate the poem in a contrived way, which shows that Al-Shanfara is either suddenly resting, or dropping dead after a long and exhausting marathon across hell. In these lines Al-Shanfara presents his reader with an image of paradise-like place where he sees himself as a male deer surrounded by flocks of female deer that look like veiled virgins:

The dust-hued mountain deer
Roamed around me,
As if they were maidens
Trailing long-trained gowns.
Toward sunset motionless they stood about me,

As among the white footed deer I were
A long horned buck, heading for the mountain peek,
Unassailable (L. 67-68).

Analysis

Al-Shanfara's early assertion that no harm on earth will befall a man traveling in the vast wilderness is not true. For the rest of the poem - lines 21 till 68- shows the falsity of the poet's initial assertion as it demonstrates the agony and extreme hardships he went through in the deadly barren desert. His life is seen, contrary to his earlier assumptions, as extremely painful and demonizing. He suffers extreme hunger until he becomes like a weakling wolf; he suffers extreme heat and extreme cold to the extent that he burns his beloved bow and arrows to heat his diminishing body; he suffers reduction in weight and size and loss of flesh to the extent that he becomes as skinny and slimy as a desert snake; and he suffers extreme reduction in his garments and cleanliness until his long hair becomes like a tail of a camel caked in urine and excrement, peppered by the desert sand and baked by the burning desert sun.

The desert is seen in the poem as a damned, horrible place. Though some of the deserts traversed by al-Shanfara are relatively inhabited with life, this does not make them any better in his eyes and the eyes of their inhabitants. In fact, the desert is not a place to be lived in but by necessity it has to be crossed. Unlike the much later European romantic descriptions of the desert[6], in the poem the desert is seen as a section of hell on earth. The creatures that inhabit it or traverse it, including wolves and people, are famished and constantly weakened by their sheer presence there. Lines 26-34, 45-49 of the poem give a detailed image of the impoverishment of body and soul of the desert fugitives. The wolves, in spite of their strength and endurance, are rendered by the barrenness of the desert into weakling creatures hardly able to drag themselves from one place to another in

---

[6] For instance, Pierre Loti, a literati traveler, views the desert through rose-tinted glasses, writes with great fondness about his experiences traveling in the Arabian deserts. In his narratives he focuses on the magical silence of the desert and the lulling pace of the camels in the middle of "pinky empty vast space" (26). Similar romantic view of the desert is well vocalized by Brosman who sees the desert as a "source of life, beauty and infinite freedom" (121). Apparently, it is such romantic notions of the desert that urged Diolé to issue a warning against picturing the desert as a convalescent home or a place to retire to: "what a misconception this is! The desert enriches only those who are already rich. It strengthens only the strong" (1956: 18).

search of what might sustain their lives. The same goes for al-Shanfara, except that his plight is more than that of his feral comrades. In addition to his weakened body, his soul is immensely tortured by his loneliness in a semi eternal vastness of aridity. Also, the desert, in the poem, is the abode of snakes, which in turn represent fear, death and evil. Al-Shanfara admits that after a prolonged stay in the desert, he becomes like the desert snake slithering unshod and in tatters under the burning sun of that horrendous place. (L. 49) He then turns into an evil marauder who remorselessly and unashamedly admits to attacking and killing men in a fiendish way:

> I widowed women
> And orphaned children,
> Then returned as I set out,
> The black night blacker still (L. 56)

Thus, in the poem, the desert becomes a symbol of evil, temporarily occupied -and not inhabited- by creatures on their way out of it. Al-Shanfara eventually got out of it. In lines 65-66, the poet gladly asserts that he finally traversed on foot a desert, as plain and barren as the back of a shield, from its beginning till its end, often a time crouching on tops of mounds to observe his pursuers if there is any. It is then, and out of the desert, that al-Shanfara had a communion with the inhabitants of the hills and mountains. His final encounter with the mountain deer whom he resembles to elegantly-clothed beautiful maidens enhances the point that life exists out side the desert.

Going through the horrendous desert on one's own costs dearly. The final scene in the poem, in which Al-Shanfara compares himself to a deer, points strongly to a radical metamorphosis in the poet's perception of himself as he has been tamed by his encounter with the reality of the desert. His reference to himself as a male deer among female deer stands in sharp contrast to his earlier boasting that his new family is made of wolves, leopards and hyenas and that he is their match and superior.[7][7] The deer in Arabic literature and in the Arab mind is a metaphor for female beauty and a symbol of elegance and playfulness, and is totally seen as the other of the previously mentioned animals. Thus, the last two lines reveal a dismaying metamorphosis in the personality of al-Shanfara from a predator to a potential victim of the desert predators. His experience in the desert has reduced the intensity of his views of the constituents of radical masculinity, and makes him akin to the men he

[7] Line 5 reads: "I have closer kin than you, /a wolf, swift and sleek, /a smooth and spotted leopard, /and a long-maned-hyena."

talks against in lines 15-18. In these lines he describes himself as a macho man. He cogently stresses the great difference between his behavior and that of domesticated and homely men of the tribe whose main concerns are their looks, coiffure and courting, and who would council and engage their wives in their affairs. Ironically, at the end of his journey he turns into what he earlier abhorred before his entrance into the desert.

Al-Shanfara's one-way journey from home, involves naturally his departure from his tribe, the fall in the crucible of physical and psychological hardships, the wondering in the form of plodding from one desert to another. Though there is no return in al-Shanfara's trip, his exit from the last desert space at the end of the poem, and his finding himself in what looks like a desert oasis with deer and mountain goats around can easily be marked as a sign of return. This return is marked by his transformation. As I have mentioned earlier, traveling through the desert changes the traveler's personality and his perception of himself. The poem is inaugurated with the traveler's expression of his intent on replacing his kinsmen with predators and his assertion of his ferocity, endurance and sense of radical masculinity. The journey into the desert eventually transforms the traveler into a deer surrounded by female deer.

Al-Shanfara's transformation perpetuates his discontinuous state of being. He starts his journey with a mindset and ends it with another. This discontinuity is shown in his emphasis, at the very beginning of the journey, on his superior moral values that egg him to leave his decadent kinsmen while he himself, in fact, is deprived of any moral values. We only realize this discontinuity when he resorts to killing other innocent men to feed on whatever he finds in their possession. He seems to be proud of his evil act by blowing it out of proportion: "I widowed women/ And orphaned children." (L. 56). Al-Shanfara blatantly takes pride in his return to the primitive. When he attacks a caravan of travelers and kills several men, he boasts about his bestiality and takes pride in the fact that the people attacked thought that their attacker is a wild ferocious beast or a demon:

> If it was one of the jinn,
> Then he is a more sinister night visitor;
> And if it was a man-
> Men do not act like that! (L. 60)

Al-Shanfara's simulation of bestiality is nothing but a strong desire to delve into a primitive stage and remain and die if necessary there.

Conclusion

To conclude, al-Shanfara's account of his journey across the desert in the company of the wild beasts proves to be very unpleasant, traumatic and has ended in the transformation of the personality of the traveler as a result of his relentless interaction with the desert. John Van Dyke succinctly and accurately describes life in the desert in a way that illuminates what befell al-Shanfara in his travels. For Van Dyke the desert is "peculiarly savage." It is a "show of teeth in bush and beast and reptile. At every turn one feels the presence of the barb and thorn, the jaw and paw, the beak and talon, the sting and the poison thereof" (56). Ibn Battutah confirms, in his travel book, the treacherous nature of the desert and cautions that "he who enters it is lost, and he who leaves it is born" (43). The final image presented by the traveler of himself as a male deer surrounded by female deer upon his emergence from the desert is problematic. In the light of Al-Shanfara's earlier affirmation of his ferocity and bestiality, his concluding lines carry the possibility of several meanings. On the one hand, his transformation from a predator to a potential hunt is a metaphor of death incurred through his interface with the desert; on the other hand, it can be seen as a humble rebirth. His implicit assertion of his sexuality by seeing the mountain deer as beautiful maidens circling him is both a jovial expression to celebrate his emerging alive from the belly of death; and an unconscious rejection of the loss of the initial vitality with which he is launched into the desert.

Works Cited

Al-Isbahani, Abu al-Faraj. *Kitab al-Aghani.* Cairo: Dar al-Sha'b, 1970.
Al-Zamakhshari, Muhammad ibn 'Umar. *A'jab al-'Ajab fi Sharh Lamiyyat al-'Arab*. Istanbul: Matba't al-Jawa'ib, 1820.
Battutah, Ibn. *The Travels of Ibn Battutah*. London: Picador, 2002.
Beeston, A. F. L. et al. *Arabic Literature to the End of Umayyad Period.* Cambridge: Cambridge Univ. Press, 1983.
Brosman, Savage. "Desert." *American Scholar* 70 (2), (2001): 111-122.
Diolé, Philippe. *Sahara Adventure*. New York: Julian Messner, 1956.
Doughty, Charles. *Travels in Arabia Deserts*. Boston: Houghton Mifflin, 1908.
Hussein, Taha. *Al Shi'r Al Jahili*. Cairo: Dar Al-Nahr Press, 1996.
Lichtenstadter, Ilse. *Introduction to Classical Arabic Literature*. New York: Twayne, 1974.
Loti, Pierre. *The Desert.* Salt Lake City: University of Utah Press, 1993.

Lyall, Charles James. *The Mufaddaliyat: An Anthology of Ancient Arabian Odes Compiled by al-Mufaddal Son of Muhammad according to the Recension and with the Commentary of Abu Muhammad al-Qasim ibn Muhammad al-Anbari*. Oxford: Clarendon Press, 1918.

Stetkevych, Suzanne Pinckney. *The Mute Immortals Speak: Pre-Islamic Poetry and the Poetics of Ritual*. New York: Cornell Univ. Press, 1993.

Van Dyke, John. *The Desert*. New York: Scribner's, 1901.

STRUGGLING TO RETURN HOME: POST-COLONIAL PERSPECTIVES ON LITHUANIAN DEPORTATION

JERILYN SAMBROOKE

Abstract

Postcolonial scholars have written extensively on how places change as the people who live in those places experience traumatic events. The place of home—and its connection to the uncanny/the unhomely—becomes a focal point for Homi Bhabha's investigation of these ideas, and this paper relies on Bhabha's work to open up a fuller discussion of how "home" can be understood as a place of change in the context of Lithuanian post-Soviet politics. The Baltic States have not usually been considered a postcolonial place (David Moore, 2001). Speaking of Eastern Europe with postcolonial language is becoming more common, but it is an emerging phenomenon, as demonstrated by Violeta Kelertas' recent publication of a collection of critical material on the topic: *Baltic Postcolonialism* (2006). My paper argues that Eastern Europe, specifically Lithuania, is a postcolonial place and that postcolonial discourse opens up possibilities for new readings of the recent—and painful—past. Referring to Dalia Grinkevičiūte's autobiographical narrative "Lithuanians by the Laptev Sea," which recounts the mass deportations of Lithuanians to Siberia by the Soviets in 1941, the paper will argue that postcolonial theory challenges common approaches to the text, which is read widely throughout Lithuania. Working with Homi Bhabha's conceptions of history in *The Location of Culture* (1994), the paper will suggest that painful experiences in Lithuania's recent past, such as the deportations of 1941, are best read as continually shaping and reshaping the present culture of Lithuania rather than read as departure points out of which the Lithuanian nation-state has been created.

Lithuania—alongside many postcommunist states—is working hard to establish an identity in the context of significant social and political shifts of the past fifteen years. Its declaration of independence from the Soviet Union in 1990 was followed quickly by its admission into the European Union (May 2004). Lithuania has been intently focused on moving westward, applying as early as January 1994 for NATO membership, and now today developments such as western shopping malls and high speed, autobahn-style highways speak to the fulfillment of this vision.

The struggle of Lithuanians to distance themselves from their Soviet history expresses itself in discussions of their literature as well.[1] Dalia Grinkevičiūtė's autobiographical account of the 1941 deportations to Siberia is a well-known text in Lithuania, still commonly read in high schools. This paper will outline the dominant reading of this text, discussing its relationship to the narrative of national identity and Lithuania's political independence. In response to growing discussions of the relevancy of post-colonial theory for literature from post-Soviet[2] countries[3], this paper will consider how the work of Homi Bhabha can be used to read Grinkevičiūtė's narrative of deportation in a new light. Rather than restricting her narrative to a recounting of past atrocities, which distances the reader from those events, Bhabha's work demonstrates how these narratives remain relevant to contemporary narratives of identity. Bhabha's discussion of how the past continually interrupts the performance of the present and his use of the term "unhomely" will be the theoretical focus. Simply stated, how can post-colonial discourse open discussions about the difficult—and recent— experiences under the Soviet regime in order to more fully understand contemporary Lithuanian identity?

---

[1] The argument of this presentation has since been developed more fully into an article, published under the title "Narratives of Identity: A Postcolonial Rereading of Lithuanians by the Laptev Sea." in *Lituanus* 54.4 (2008): 8-26. The material presented here is published with the permission of the editors of *Lituanus*.

[2] Post-Soviet and postcommunist will be used interchangeably, as they are in much of the literature in the field.

[3] For more detailed discussion, see Adrian Otoiu "An Exercise in Fictional Liminality: the Postcolonial, the Postcommunist, and Romania's Threshold Generation." Comparative Studies of South Asia, Africa and the Middle East. 23:1&2 (2003): 87-105. Otoiu highlights the differences between communism and capitalism, the ideologies that created different forms of colonization, and also discusses why the discipline of postcommunist studies has been slower to emerge than postcolonial studies (language barriers, difficult of dealing with recent history, etc.). Kelertas also discusses the differences between postcommunist and postcolonial theory in her introduction to *Baltic Postcolonialism*.

The problematic relationship between postcommunism and post-colonialism must be acknowledged, however, before advancing the larger argument. Resistance to including post-Soviet states in the umbrella group of post-colonial states has come from both inside post-Soviet communities and from inside post-colonial studies.[4] [3] Violeta Kelertas recognizes the resistance to post-colonialism in the introduction to her edited collection *Baltic Postcolonialism* (2006), and the publication of the volume is a clear argument against the false division of the areas of study. In various ways, each of the submissions to this collection of essays addresses this same resistance, and Almantas Samalavičius, professor at Vilnius Gediminas Technical University, strongly asserts that resistance to using post-colonial theory comes directly from the academic community in post-Soviet states:

> Efforts to synchronize its [Lithuanian academic criticism] orientation with contemporary trajectories of analysis [i.e. post-colonialism] are judged by the institutional critics as polemics of little merit or ideologically dangerous heresies. (410-411)

This "internal" resistance to applications of post-colonial theory is further compounded by the reluctance of the post-colonial academic community to include post-Soviet states in theoretical dialogue. This question was brought to the fore in 2001 when PMLA published David Moore's article entitled "Is the Post in Postcolonial the Post in Post-Soviet?" In his article, Moore questions why the academic community recognizes an astonishing variety of countries as post-colonial in some manner but exempts post-Soviet states from this discourse. In his project on post-colonial analysis of Romanian literature, Adrian Otoiu comments on difficulties he has had on gaining a hearing for his work at post-colonial academic conferences.

In the face of such resistance to considering post-Soviet states as post-colonial, Samalavičius insists that Lithuania must be understood as post-colonial:

[4] A conclusive argument proving that Lithuania should be defined as a post-colonial nation-state lies beyond the scope of this paper. Several scholars referenced in this paper address this question, however. David Moore, for example, argues that all regions or nations could fruitfully be considered as "postcolonial," just as each nation can be analyzed using Marxist theories. Almost all people have been conquered and/or have conquered at various moments in history, and to draw time frames around the application of postcolonial theories is unnecessarily limiting.

> [...] if the past is not understood as a protracted and conscious colonization executed by an exterior power (i.e. the Soviet Empire), it is impossible to adequately evaluate the habits of thought and social activity, as well as the pathologies of self-identification formed during that period, which, though they were not as obvious as those in the societies of the so-called 'Third World,' left their imprint not only in Lithuania but in the other Baltic states as well (Kelertas 411).

Samalavičius is careful here to define colonization as "protracted and conscious," arguing that only when Lithuania sees its history in such a light will Lithuania be able to move forward. The implied distinction that Samalavičius makes here between colonization and occupation often rises to the surface during discussions in which "postcolonial" is deemed inapplicable to post-Soviet states.[5] By using the term "occupation" rather than "colonization," Lithuanians can point to a clear end to the "occupation," namely the declaration of independence in March 1990, thereby distinguishing clearly the political processes of the occupation from the cultural processes of Lithuanian identity. "Occupation" implies that there was only a formal political engagement of the two cultures in which one culture is suppressed completely against its will. The term post-colonial, rather, implies a depth of social influence where cultural patterns, "habits of thought and social activity," as Samalavičius phrases it, are influenced. The resistance to the term "colonization"—and by extension the resistance to the term "post-colonialism"—that Samalavičius refers to is what this paper will investigate as it relates to the deportation text by Grinkevičiūtė.

Introduction to Grinkevičiūtė's Text

Dalia Grinkevičiūtė's entire family was exiled to Siberia in June, 1941 during the early years of World War II. The Stalin-Hitler, or Non-agression pact, of 1939[6] made provisions for the expansion of Soviet

---

[5] Many of my students in Eastern Europe have consistently relied on this distinction to argue that postcolonial theory cannot be applied to post-Soviet states such as Lithuania, Latvia, or Ukraine.

[6] August 23, 1939, Soviet foreign minister Vyacheslav Molotov and the German foreign minister Joachim von Ribbentrop signed the Molotov-Ribbentrop pact, also known as the Hitler-Stalin pact (Molotov-Ribbentrop pact, Nazi-Soviet pact). Although the treaty was officially labeled as the Treaty of Nonaggression between Germany and the Union of Soviet Socialist Republics, the pact included a secret protocol, in which the independent countries of Finland, Estonia, Latvia, Lithuania, Poland and Romania were divided into spheres of interest. This secret protocol

influence into Lithuania. When the Soviets invaded in 1940, they planned large deportations as an attempt to establish and maintain authority in the region. In June 1941, these deportations began and Grinkevičiūtė's family was among the 17 000 Lithuanians who were sent to various concentration camps across Russia (Kiaupa 377). The deportations were forced to stop when Hitler broke the pact with Stalin and invaded Lithuania on June 21, 1941.

Grinkevičiūtė's father was sent to a concentration camp in the Northern Urals and died after two years of starvation and overwork. She, along with her mother and brother, were sent to a collective farm, then to a work camp in the far north, and then finally to a an uninhabited island 800 kilometres beyond the polar circle (Grinkevičiūtė 62-3). From 1941 to 1949 Trofimovsk Island was used as a concentration camp, and few people survived. Grinkevičiūtė's memoirs recount the horrors of life at Trofimovsk.

The memoirs themselves also have a long and complicated history, and a brief outline of their publication is important in contrasting a post-colonial reading of the text with the more dominant, nationalistic reading of the text. The earliest version of the memoirs was written around 1950 but was buried in a glass jar at the time. They were discovered in 1991, published in Lithuanian in 1997 and in English in 2002. Grinkevičiūtė continued to write after this initial version was completed 1950, however, and a later version, which is shorter and more focused, has been more widely circulated. This later version was published illegally in Moscow in 1979 and from there smuggled abroad and translated. While illegal copies could be found in private homes in Lithuania, it was not until 1988 that it was published in Lithuanian.

The political and social context of this first publication has strongly influenced the reading of Grinkevičiūtė's memoirs, both in 1988 and still today. In 1988, Lithuania was in a turbulent transition, moving towards its official declaration of independence in 1990. Those who published the memoirs were working to unite the Lithuanian public around a common narrative that would encourage identification with the spirit of independence. With this goal in mind, the manuscript was not only selected and published, but also titled. Grinkevičiūtė had never titled her manuscript, and since the text was published posthumously in 1988 (she

---

explicitly assumed territorial and political rearrangements in the areas of these countries. This resulted in a new occupation campaign against Lithuania on June 15, 1940, when the Soviet government began moving troops into all regional centers and strategic locations that resulted in Lithuanian President A. Smetona withdrawing to Germany (Gostautas).

died in 1987), it was the publishers who chose the title Lithuanians by the Laptev Sea (Everatt in Grinkevičiūtė 178), now quite well-known in Lithuania. Including "Lithuanians" in the title, argues literary critic Jūra Avižienis, elicited a particular reading of the text, one that served to nationalize Grinkevičiūtė's experiences:

> The title nationalized the experiences [...] and elicited a particular kind of reading during the years leading to and immediately following Lithuanian independence [...]: it demanded that the memoirs be read as a statement about how Lithuanians as a national group were treated and impacted by Soviet occupation, and it denounced Soviet incorporation by Lithuanians as a national entity. (193)

By encouraging the Lithuanian reader to identify with the text along national lines, the publishers contributed significantly to the larger narrative of independence forming in Lithuania at the time. The deportations were framed as an example of Soviet treatment from which all Lithuanians should now work to free themselves.

Recent criticism surrounding and comments about this text continue to echo this emphasis on national identity. In 2002, an English translation of the full-length memoirs (the version which was buried in 1950) was published with an introduction by a prominent politician, Vytautas Landsbergis. Landsbergis was a key leader in the Lithuanian Independence Movement and later became the first head of state of independent Lithuania. Currently, he represents Lithuania in the European Parliament. Both the editor's choice to have such a prominent political figure introduce the text and his specific comments illustrate that the nationalized reading of the text still persists. The text continues to be relied on to provide a point of collective identification, as it is read as an account of what happened to us.

In the reading of Grinkevičiūtė's memoirs that Landsbergis offers, the text is seen as offering the true version of what happened. The text is primarily an eye-witness account, given social and political prominence because it legitimately challenges any previous Soviet suppression of the deportations to Siberia. Landsbergis begins his introduction with a bold, direct statement about Grinkevičiūtė's intent to document the truth: "Dalia Grinkevičiūtė's mission was to testify about what happened" (5). To testify about what happened means, for Landsbergis, to bring into the open what atrocities the Soviet regime committed against the Lithuanian deportees.

Secondly, Landsbergis encourages the reader to interpret the text following his introduction as representative of Lithuanian courage—an ability to survive in the face of unimaginable suffering. He frequently draws attention specifically to the national identity of the deportees, encouraging his readers to wonder at the admirable struggle for survival and hope of these Lithuanian deportees: "There were a great number of deportees in the north. The songs of Lithuanians could be heard along the banks of the Angara in Siberia…. 'Let me go back to my homeland' became the anthem of the deportees" (6). National song is a strong tie in any culture, and Landsbergis is deliberately encouraging his readers to see the emotional attachment of these people to the nation of Lithuania. He is also encouraging his Lithuanian readers to identify emotionally with these Lithuanian deportees. He reminds his readers that they share these songs with these deportees.

This reading of the narrative, however, implies a cause and effect logic of history that is problematic. While I would not go so far as to say that Landsbergis is incorrect in his reading of the text, the focus on the historical and national nature of the deportation results in an incomplete reading of the text. The text is seen to give a departure point from which Lithuanians now move. The text becomes a "cause," representing the deportations and violence of the Soviet regime, against which contemporary Lithuanians are encouraged to respond. By supporting an independent, free Lithuania, in which laws protect those who were formerly persecuted, one is seen as responding appropriately to the violence of the past. If the deportation is the "cause," then a free Lithuania is the "effect." Grinkevičiūtė's text, then, is useful in as far as it furthers the political agenda of an independent Lithuania, but the experience of deportation is then subjected to this narrowly defined reading. All details of the experience that do not relate to this political narrative are forgotten, or at least deemed less significant. The logic of cause and effect simplifies both the experience of deportation and the role of the narratives in the Lithuanian community. Reading the memoirs as a statement of national identity—as done both at the time of their publication in 1988 and in current discussions— encourages Lithuanian identity to be developed by using the Soviet past as a departure point along a chronological timeline towards independence.

Turn to Post-colonial Theory

In an effort to grasp more fully how Grinkevičiūtė's text can be read to go beyond this cause and effect view of history, I will turn to post-colonial theory. In the introduction to Baltic Postcolonialism, Violeta Kelertas refers to Bill Ashcroft's work, specifically Post-colonial

Transformation (2001), to define a post-colonial approach that is specifically relevant to Lithuania:

Ashcroft now describes "'post-colonial' as representing a form of talk rather than a form of experience....If we see post-colonial discourse in the Foucauldian sense as a system of knowledge of colonized societies, a space of enunciation" in which "the experience of oppression, of invasion, of domination" can be expressed, discussed and analyzed from a fresh perspective, then we can begin to comprehend, validate and honour all the suffering and give it a name (Kelertas 8). If the post-colonial is a "form of talk" rather than a "form of experience," then attention must be paid to how communities interpret—talk about—the narratives of their past. Homi Bhabha's work provides a framework in which new readings of Grinkevičiūtė's memoirs can be explored.

Bhabha argues in *The Location of Culture* that "post" colonial is "beyond" rather than "after" colonial. He develops this conception of the past in relationship to cultural works that emerge from the periphery. He explains:

The borderline work of culture demands an encounter with 'newness' that is not part of the continuum of past and present....Such art does not merely recall the past as social cause or aesthetic precedent; it renews the past, refiguring it as a contingent 'in-between' space, that innovates and interrupts the performance of the present. (Bhabha [1993] 10) If the past is not simply the social cause for a text but rather works to interrupt the performance of the present, then it becomes possible to interpret Grinkevičiūtė's memoirs as more than a statement defining Lithuanian identity in opposition to Soviet mistreatment. She is calling the past—the experiences of deportation—into the present, providing insight for the reader on how these experiences have shaped her as the author/narrator (however problematic that relationship may be). The narrative voice and the structure of the text create the effect of also posing this question of identity to the reader. By drawing the past into the present in this manner, the text urges the reader to acknowledge the impossibility of ever moving on from the past.

Running parallel to the past interrupting the present in Grinkevičiūtė's memoirs is the foreign interrupting the familiar, or the world interrupting the home. Bhabha has developed the term "unhomely" in his work to discuss how the neat division between what is unfamiliar, or "unhomely" and what is familiar or "homely" is often not neatly divided. In his essay "The World in the Home" (1992), he defines the unhomely moment as "the shock of recognition of the world-in-the-home, the home-in-the-world" (141). The word "unhomely" is a deliberately awkward translation of the original German "unheimlich," which has a tidy opposite of "heimlich," or homely. In developing these concepts, Bhabha suggests that this space between the "heimlich" (homely) and "unheimlich" (unhomely) is a post-colonial place, a space in which one sees how a person's identity is a complex mixture of what is foreign and what is familiar. Layered on top of this tension between the foreign and familiar is the second, more common, translation of "unheimlich"—"uncanny," clearly echoing the work of Sigmund Freud. Bhabha is interested in applying the idea of the uncanny to the post-colonial experience. David Huddart summarizes how Bhabha applies this concept to the understanding of identity:

If the uncanny is homely, what is close to home, it none the less has a tendency to morph into the profoundly unfamiliar, the unhomely, which estranges us from what we thought was most properly our own. Alienation would usually be thought of as a problem, but if it is something which is part of all experience, and is even something which might inspire us to re-evaluate our identities, then we can understand it as an opportunity. The uncanny, in other words, opens a space for us to consider how we have come to be who we are. (83) In the act of writing about her deportation, Grinkevičiūtė brings the foreign into the home, the pain of deportation home to the Lithuanian reader. In acknowledging the uncanny effect of the structure of the narrative, the critic and the reader can then pause to consider how Grinkevičiūtė's text does not simply document the deportation experience but rather can consider how the text opens a space to consider how one has come to be who one is. The narrative—and the experiences of deportation—reenter the present, not as a text which justifies Lithuania's independence, but rather as a text that continually asks its reader to consider how these experiences influence the reader's current identity.

Analysis: Grinkevičiūtė's Memoirs

There are two specific aspects of this work—the narrative voice and the stylistic tone—that effectively address the reader in a direct manner, and, by producing this effect, these aspects of the work demonstrate how the events of the past "interrupt the performance of the present" and also create an "unhomely" moment for the reader. These concepts of Homi Bhabha's, when applied to a reading of Grinkevičiūtė's text, encourage the community which reads these memoirs to reconsider the role of history in the present. If these events are considered to be precursors to or justifications for the establishment of independent Lithuania, then the impact of these events will only be partially understood by that community. If post-colonial theory can be used to demonstrate that these events continue to shape this community, then the vision of Ashcroft and Kelertas, referred to earlier, to create a "space of enunciation," where these experiences are continually given voice, will become a reality.

Narrative Voice

It is specifically Grinkevičiūtė's narrative voice that opens the possibility for reading the text as an expression of how the past interrupts the present. As an autobiographical piece of writing, the narrator, the main character, and the author all claim to be the same individual and yet are distanced from one another as they exist in different moments of time. By moving throughout the text between the "I" who is the autobiographical writer and the "I" who is the young, exiled girl of fourteen, Grinkevičiūtė creates a gap. The past is not simply a "cause" for the text's plot, a series of preceding events, but clearly the "I" who is the author has an identity that comes out of these past events. While this could be seen as a basic feature of any autobiographical writing, Grinkevičiūtė uses the narrative voice to create the effect of inviting the reader into a similar process of questioning how the past influences the identity of the reader in the present. When the reader becomes involved in this questioning of his/her identity, then the past begins to function in a manner similar to that of which Bhabha speaks.

Throughout the text are brief statements in the present tense that interrupt the flow of the text. These interruptions in the present tense have two effects. First, they push the reader to acknowledge that the experiences of the past continue to define the present for the narrator. Secondly, these moments invite the reader – who now occupies this present space with the narrator – to see how the past defines the reader's own identity in the present.

One of these changes in tense occurs during the description of the awful polar nights of constant darkness, during which many people died. Due to scurvy, many deportees could not stand even to relieve themselves and were forced to lie in their own excrement. Lice infested even their eyebrows and eyelashes. "The end seemed imminent" (76) does not seem overly dramatic as a summary of the situation. Into this hopelessness arrived Doctor Samodurov. For a month he worked to help the prisoners simply survive by negotiating food rations, opening the bath house, and using the disinfection chamber. It was a turning point in the lives of those deportees: "After a month Doctor Samodurov left. We heard that he had been killed at the front, but then, maybe that wasn't true? We all bow to you, Doctor Samodurov" (76). Who, exactly, is bowing to the doctor? Implied is the "we" who were saved by his work. However, the present tense complicates the use of "we" in that way, since many of that group would have already died by the time of writing. Perhaps Grinkevičiūtė means simply those prisoners who were still alive as she was. I would like to suggest a broader interpretation; she and her readers bow to the doctor. All those who are thankful for what he did – the "we" who join her in the act of telling and of reading the story – these are the ones who bow. Grinkevičiūtė seems to invite the reader of her text into this process of identification with the past.

Even if it is argued that the reader is included in the "we" who acknowledge the doctor, however, this act of bowing to the doctor can still be read in a nationalistic sense, encouraging the Lithuanian reader to identify sympathetically with the deportees along ethnic lines. This collective sympathy encourages Lithuanians to acknowledge the horrors of the past and respond by creating or maintaining an independent Lithuania. It may also be read in a second manner, however, that is more consistent with how the other uses of the present tense in the narrative. Rather than "acknowledging" the past, which allows a person to see the event as it "truly was" and then act on it, the reader is being asked to see how these experiences continue to effect the narrator and, by extension, the reader. One cannot "move on" from these events. They haunt the writer, and the narrative is less focused on "telling the truth" of what happened (did the doctor really die?) than it is about asking the reader to allow himself/herself to be haunted by these events as well.

The second example of changing tense in the narrative voice is at the end of the text. The narrator tells in the present tense of the burden she feels for these people: "The dead continue to live in my heart. Many years have passed but I can still see them….It is my duty to tell their story" (79). In her inability to push these people from her mind, she recognizes that she will never move beyond the past but rather must continually retell it, bringing it into the present:

> Now Trofimovsk Island is once more empty and uninhabited. During storms the waves of the Laptev Sea strike the shore with great force and steadily erode it.
>
> In 1949, when the guards took the last of the remaining exiles to other places to fish for the state, the waves had already begun to destroy the edge of the common grave. It began to crumble. I am absolutely certain that by now the waves of the Laptev Sea have long since washed away all the corpses.

> I wonder which seas and oceans they wander, searching for the way back to their homeland. (80)

For those who died at Trofimovsk, their only way back to their homeland is through Grinkevičiūtė as a writer. Narrative always functions in the present and demands that the storyteller and the audience articulate their current identities in relationship to the events that are being told. As Grinkevičiūtė invokes the presence of the wandering dead, she finds herself spanning the distance between Lithuania and Trofimovsk Island. The "I" who knows those people wanders with them, and recognizes that her identity is only able to be understood in the context of this wandering. By narrating these events in the present, she pushes her audience, her readers, not only to "react against" these events in the past, but, more significantly, to see how this wandering among the dead is shaping their identities in the present.

Narrative Tone

The second aspect of the text that is important to consider in this alternative reading of the text is the tone of the narrative. The narrative is told in the first person, past tense in a very restrained tone. Grinkevičiūtė includes minimal emotional response to the horrific events that she is recounting, and there is a level of detail in the text that creates a tone of documentation. An obvious explanation of this kind of detail is the role of this text in recording and publicizing the horrible Soviet treatment of Lithuanian deportees: what exactly occurred, who was involved, what was the outcome of the situation? These details most commonly include names or references to indicate identity. For example, in recounting a story of fifteen young men who tried to escape and died during their attempt, she takes care to share all she knows: "Of the fifteen, I remember only one man's last name: Zobiela" (65). It is also common to read a name and an association with a profession or hometown: "In our barrack Atkočaitienė, a printer's wife from Kaunas, lay dead on the planks beside the living for several days" (71). During a listing of individual deaths, Grinkevičiūtė tells of a professor: "The dean of the Lithuanian Agricultural Academy, Professor Vilkaitis, was assigned guard duty. Exhausted by starvation, the professor collapsed on top of some barrels and died. Out of respect for the professor, who was well known even beyond Lithuania's borders, the people built him a coffin" (72). These three examples are a few of numerous similar statements throughout the text.

Grinkevičiūtė was writing for an audience who could have known these people. As the reader encounters names of so many individuals and families, it begins to seem possible that the reader will encounter a familiar name. As this expectation mounts, the reader's approach to the narrative easily shifts, as the reader awaits a familiar name, or at least a region, a city, or an institution—something that would connect the reader to the people in the narrative. In a small community of fewer than 3.5 million, this is not an implausible expectation. By structuring the narrative in this way, Grinkevičiūtė encourages an engagement with the text that differs substantially from a reader's engagement with a narrative focused on keeping records. Grinkevičiūtė changes the reading experience by inviting a personal connection to the incomprehensible events that she relates. The finding of the familiar—the ability to recognize one's home (or neighbours, or city, or university)—in the very unfamiliar narrative of deportation, starvation, and death is the discovery of the uncanny, "the unhomely," in the narrative. In that moment of recognition of the reader's home in the midst of the suffering of Trofimovsk Island, the reader is forced to pause and consider how that moment changes "home." The world, the foreign, has interrupted the home, and the reader must now respond. This is no longer a narrative that only documents the horrors of Siberia, asking people to respond so that such horrors will never be repeated, but the narrative functions in a manner that continually draws the reader into the question of "who am I" in light of my home connecting to that world.

Conclusion

The reluctance to apply post-colonial theory to Lithuanian texts is not only academic in nature but also personal. Lithuanians are looking forward. Entry into the European Union and NATO have opened up new possibilities but also posed new challenges. While the suffering endured by Lithuanians of the past is important, many people simply want to move on. They want to look to a new future with fresh possibilities.

To read one's past as a departure point for a new life, a life that is detached from the suffering of the past, is to simplify, on one hand, the experiences of the past, and also to simplify the effect of those experiences on one's current identity. By challenging the cause and effect view of history that lies behind the dominant, nationalistic reading of Grinkevičiūtė's memoirs, a space is opened in which to acknowledge with Grinkevičiūtė that one cannot simply turn one's back on the past and move forward. The text challenges its readers to allow the past to destabilize the present and to allow foreign events to demand a local, personal response. The courage to acknowledge the past in the present is at the core of post-colonial discourse, and Grinkevičiūtė had this courage. Her memoirs request that same courage of her readers.

Works Cited

Avižienis, Jūra. "Learning to Curse in Russian: Mimicry in Siberian Exile." In Kelertas. 187-197.

—. "Mediated and Unmediated Access to the Past: Assessing the Memoir as Literary Genre." Journal of Baltic Studies. 36.1 (2005) 39-50.

Bhabha, Homi. *The Location of Culture*. London: Routledge, 1994.

—. "The World and the Home." Social Text. 31 (1992) 141-153.

Edwards, Brent Hayes. "The Genres of Postcolonialism." Social Text 78. 22.1 (2004) 1-15.

Gostautas, Richard. "Exile and Imprisonment Locations of the Residents of Lithuania (1941)." Lithuanian Genealogy Database. Richard Gostautas. 20 May 2006. <http://www.lithuaniangenealogy.org/databases/lithuania/exile/index.html?letter=E>

Grinkevičiūtė, Dalia. "Lithuanians by the Laptev Sea: An Excerpt From a Memoir." Trans. Laima Sruoginis. 60-80. In Sruoginis.

Huddart, David. *Homi K. Bhabha*. London: Routledge, 2006.

Kelertas, Violeta, Ed. *Baltic Postcolonialism*. Amsterdam: Rodopi, 2006.

Kiaupa, Zigmas. *The History of Lithuania*. Trans. S.C. Rowell, Jonathan Smith, and Vida Urbonaičius. Ed. S.C. Rowell. Baltos Lankos: Vilnius, 2002.

Landsbergis, Vytautas. "Introduction: A Piercing Light." A Stolen Youth, A Stolen Homeland: Memoirs.

Otoiu, Adrian. "An Exercise in Fictional Liminality: the Postcolonial, the Postcommunist, and Romania's Threshold Generation." Comparative Studies of South Asia, Africa, and the Middle East. 23: 1 and 2 (2003): 87-105.

Samalavičius, Almantas. "Lithuanian Prose and Decolonization: Rediscovery of the Body." Trans. Violeta Kelertas. 409-427. In Kelertas.

Sruoginis, Laima, Ed. *The Earth Remains: An Anthology of Contemporary Lithuanian Prose*. Vilnius: Tyto alba, 2002.

ACROSS THE ZODIAC: LEXICOGRAPHICAL TRAVELS THROUGH SPACE, TIME, INFINITY AND BEYOND

JAMES LAMBERT

Abstract

Across the Zodiac: A Historical Dictionary of Science Fiction is an exhaustive treatment of the language of Science Fiction in the form of a dictionary on historical principles. It attempts to trace the origin, history and sense development of those words, or senses of words, that either arose in the literature of science fiction, or that have special significance in the genre. Each entry not only provides the usual lexicographical information of definition, part of speech, etymology, pronunciation, etc., but also allows the reader complete access to the source and text of the citations upon which the sense differentiation and definitions are based. The venerable age of many of the terms and phrases now recognised as SF standards, or even clichés, are frequently surprising. By tracing the language of science fiction from its earliest inception, and by delving into its precursors, many of the major conceptual shifts regarding the place of human beings in the world, and the universe, are revealed. Starting from a pre-science view of a flat Earth under a heavenly abode of God(s), humanity through science and science fiction, has embarked on an ever-expanding journey through space and time and along the literary way a metamorphosis of self-perception has inevitably taken place. However, this is not all, nor has the journey ended, for science fiction deals with much more than mere space travel. Notions of human development along inseparably entwined threads of genetic mutation, physical and psychic evolution, and technological advancement, form a tangled skein of metamorphic possibilities.

Introduction

The creation of neologisms is an open playing field; anyone can invent words or new usages at will. However, language as a social tool is apt to be applied only when there is a need. The arrival of a new word on the language block signals the genesis of a conceptual novelty. Certainly complex ideas can and often are generated long before they become associated with a particular lexeme, but the settling down of an idea into a word of the language is a signification of that notion's debut into the wider discourse. Dictionaries on historical principals, tracing as they do the earliest attestations of new words and senses, are an under-utilised resource in literary and social research. Work on a comprehensive historical dictionary of science fiction, with the working title *Across the Zodiac*, provides insight into the ongoing development of the conception of humanity's place, and even form, in the universe.

Part I: Place

Science Fiction Origins and the Father of Science Fiction

Many have attempted to provide a starting point for science fiction and in doing so generally opt for any of a number of famous authors or notable works (Kress), including, Jules Verne, H.G. Wells, Mary Shelley's *Frankenstein*, Swift's *Gulliver's Travels*, even going as far back as Lucian's *True History* (Georgiadou and Larmour). However, these texts were written outside the genre, or rather, before the genre had formed, and were without parallel in their times, and are therefore better viewed as precursors to science fiction, or proto-science fiction. Precursors aside, the type of science fiction story that is typical of the genre, known as "genre science fiction" in the world of science fiction criticism (Prucher 78),[1] first began to be written in the early 1900s and this literary phenomenon is largely attributable to the efforts of Hugo Gernsback, the oft-labelled

[1] The list of science fiction subgenres is surprisingly well populated, including: *Hard Science Fiction, Soft Science Fiction, Social Science Fiction, Cyberpunk, Biopunk, Steampunk, New Wave Science Fiction, Golden Age Science Fiction, Pulp Science Fiction, Space Opera, Science Fantasy, Speculative Fiction,* and more. At any rate, all of these lay within the scope of the umbrella term *genre science fiction*.

"Father of Science Fiction,"[2] after whom science fiction's most prestigious award, the Hugo, is named.

In 1908 Gernsback started *Modern Electrics*, and then in 1913, *The Electrical Experimenter*, both magazines aimed at amateur radio and electrics enthusiasts. These magazines "carried how-to articles, information on new radio patents, and news of radio and radio operators across North America [and] Gernsback was publisher, editor, chief writer (and often ghost-writer), and also did layout and sold advertising" (Banks). In other words, the magazines were a one-man-show. It was in *Modern Electrics* and *The Electrical Experimenter* that science fiction, as we know it today, was born. Gernsback, following up on the realisation that his readers were science-minded but not trained scientists, had the idea that purely fictional stories about science might be able to give real scientists a potential insight for possible experimentation, research and development. Gernsback initially led the way by writing fictional science stories himself, starting in 1911,[3] and by 1916 was encouraging his readers to take up the pen. He initially called this new type of fiction *scientifiction*, a portmanteau word derived from blending *scientific* and *fiction* (Gernsback, 1916), though he later renamed it to the more familiar, and easier to pronounce[4] *science fiction* (Gernsback, 1927). So popular was this type of fiction amongst his readers that in 1926 Gernsback decided to start a magazine wholly devoted to science fiction, and thus began *Amazing Stories* which is still publishing science fiction today over eighty years later.[5]

---

[2] There are a number of other 'Fathers' of Science Fiction, including Jules Verne, H.G. Wells, and John W. Campbell.

[3] These were serialized in *Modern Electrics* and then in 1925 published as a novel *Ralph 124C 41+: A Romance of the Year 2660* (Boston: The Stratford Co.). The "124C 41" of the title is a pun on "one to foresee for one." Currently a copy of the first edition with rare dust jacket will set you back at a minimum of roughly US $500 and a copy signed by Gernsback in excess of US $30,000! AddALL website. Accessed 25 May 2007. (www.addall.com).

[4] Prucher (177) gives the pronunciation /ˌsaɪəntɪ'fɪkʃən/, which maintains the primary stress of *fiction* /'fɪkʃən/; another alternative would be /ˌsaɪən'tɪfɪkʃən/, in order to maintain the primary stress of *scientific* /saɪən'tɪfɪk/; a third possibility, more easily rolling off the tongue, would be /saɪ'ɛntɪˌfɪkʃən/, which preserves the stress and pronunciation of *science*. Which ever way, the pronunciation of *scientifiction* is lingually problematic.

[5] Justine Larbalestier credits Gernsback with the invention of science fiction arguing that "Science Fiction as a community, and certainly science fiction as a publishing category, began in the United States in 1926 with the first English language science fiction magazine, *Amazing Stories*" (15).

The Setting: Space and the Plurality of Worlds

Fantasies about travel into space have a long history, reaching back to mythic times with Icarus and Daedalus. In the seventeenth century Francis Godwin's astronaut Domingo Gonsales was conveyed to the Moon by a flock of geese, and Cyrano de Bergerac employed evaporating dew to effect his lunar excursion (Pizor), while in the eighteenth century Voltaire's Sirian,[6] Micromegas, travelled sometimes via sunbeam, sometimes via comet (Smollett). The late nineteenth century saw proto-science fiction writers dealing with the topic of space travel more scientifically, and in a more conventional space craft, with the publication of such works as Percy Greg's *Across the Zodiac*, in 1880, and John Jacob Astor's *A Journey in Other Worlds*, in 1894.

However, the very concept of Space, as opposed to the Earth, commonplace as it is to us today, is itself somewhat of a newcomer in the history of human ideas. In the pre-Copernican and pre-Galilean age, Earth was contrapuntal not to Space but rather to Heaven, the abode of the gods. Stars were fixed points of light in the 'firmament' (not 'space'), the sun, moon and planets were wandering stars[7], and all heavenly bodies were associated with a god or goddess. The very notion that planets themselves were analogues of our world, that they were spherical bodies of roughly similar size revolving around the Sun, and that they might therefore be inhabited by living creatures as our own globe, shook the very foundations of knowledge. Despite early attempts to suppress these heretical ideas, in time speculation abounded. Tracing first occurrences of words, or new senses of existing words, provides an insight to the origin, or the starting point of acceptance, for new concepts. The term *world* is a common Teutonic lexeme that predates the arrival of the Angles, Saxons and Jutes in England. Since that time until the seventeenth century, a period of some eight centuries, it was only used to refer to the planet Earth; it is first used to signify other planets in the solar system in 1638 (Godwin 10).[8] Similarly, the astronomical sense of the word *space* is also a modern English development, dating only back to 1667 (*OED Online*).

A natural flow-on from the acceptance of the planets as other worlds, as other places, was the inevitable question: Is there life on other planets? Within the science fiction genre the resounding answer has always been

---

[6] An inhabitant of the star system Sirius.

[7] Our modern English word *planet* is ultimately derived from Ancient Greek *πλάνητες* the 'wanderers', or *πλάνητες 'αστέρες* the 'wandering stars'.

[8] This evidence antedates the earliest record (1713) in the *Oxford English Dictionary* (hereafter *OED*).

Yes! Travelling in order from the Sun outward, science fiction is populated with *Mercurians*, *Venusians*, *Martians*, *Jovians*, *Saturnians*, *Uranians*, *Neptunians* and *Plutonians*. Not to forget the *Selenites*[9] or *Lunarians* of the Moon, nor the *Solarians* inhabiting the Sun. By attempting to trace the earliest examples of these words in English literature[10] we can get a picture of the venerable age of speculation about planetary beings other than humans. The date of earliest attestation for each of these terms, sorted chronologically, are:

1645 Selenite (Howell, III. ix. 18)
1698 Mercurian (Huygens 106)
1698 Saturnian (Huygens 123)
1708 Lunarian (*British Apollo*)
1725 Jovian (Harris 127)
1839 Uranian[11] (*Fantastical* 145)
1839 Venerean (*Fantastical* 27)
1854 Neptunian[12] (Brooman 488)
1874 Martian (Prucher s.v.)
1874 Venusian (Blair 130)
1931 Plutonian[13] (Coblentz 323)
1932 Solarian (Campbell 136)

These new words to the English language are examples of systematic word formation, appending the adjectival and noun suffix *–ian* to the name of the planet (*Mercury* + *–ian*) or a representative Latinate stem (*Jovis* + *–ian*), and could have been coined at any time since the Middle English period. However, they were unneeded in the English language until the increase in astronomical knowledge and the subsequent speculation about life on other planets required them.

With the exception of *Plutonian* and *Solarian*, both dating from the 1930s, hence post-Gernsback, these terms for inhabitants of other worlds do not first appear in science fiction proper, and for the most part there was no explanation of the imaginary inhabitants' biology, culture or technology, no interaction with them, no drama, no invasions, no

---

[9] From Greek *σεληνη* the moon.

[10] The earliest attestations provided in this paper frequently antedate those found in the *Oxford English Dictionary* (whether the second edition of 1989, the *Additions Series* 1993–1997, or *OED Online*) and Jeff Prucher's recently published *Brave New Words: The Oxford Dictionary of Science Fiction*, 2007.

[11] Uranus was discovered by astronomers in 1781.

[12] Neptune was discovered by astronomers in 1846.

[13] Pluto was discovered by astronomers in 1930.

interplanetary alliances or galactic empires, that are typical of twentieth century science fiction. Instead these terms began their lives in the then-insipient genre now known as Popular Science – scientific material written for popular audiences. In terms of astronomy and its relation to early science fiction vocabulary the seminal work was Bernard le Bovier de Fontenelle's *Entretiens sur la Pluralité des Mondes* (*Conversations on the Plurality of Worlds*) of 1686. This work was essentially an explication of celestial mechanics offered through the medium of nightly conversations between a young male astronomer and a keenly interested young woman. It was first translated into English in 1688 by no less a literary light than Aphra Behn and spawned a succession of copycat books which continued well into the nineteenth century. In these texts the inhabitants of other planets, when mentioned at all, which was not often, were posited merely to serve as a point of view in order to explain some point of celestial mechanics: "*Jupiter* performs his Gyrations in the Space of ten Hours; and therefore a *Jovian*, or an inhabitant of *Jupiter* will see the whole Heavens, and even our *Earth* together with the *Sun*, to have a rapid Motion round his Body in the Space of ten Hours" (Keill 73). Though as time went on there was more and more speculation about the possibility of life on other planets and the form that life would take, as in this 1760 translation of Fontenelle: "The optic nerves of the Mercurians must not be so delicately tender; they could not possibly support the glare of light" (373). Also, learned debate raged about whether or not life solely existed on Earth, and how Biblical teachings illuminated this point (Whewell; Brewster; Proctor).

To Boldly Go: From Inter*planetary* to Inter*stellar* to Inter*galactic* Space

The next step outwards in terms of place and expanding consciousness is beyond the Solar System. In science fiction this outward movement was made possible for humans by the "invention" of various super-technologies such as *hyperspace travel* in 1928 (Olsen 533), *space warp* in 1935 (Schachner 101), and *FTL* or *faster-than-light drive* in 1936 (Campbell 17). Through these technologies humanity was able to contact and interact with the inhabitants of other star systems, such as Alpha Centauri, Antares, Betelgeuse, Rigel, Sirius, and Vega, and neighbouring galaxies, such as the Andromeda galaxy:

1930 Andromedan (Smith 17)
1931 Antarean (Smith 132)

1932 Sirian (Campbell 84)
1935 Centaurian (Leinster 234)
1937 Alpha Centaurian (Binder 90)
1940 Rigellian (Asimov 14)
1940 Betelgeusian (Asimov 32)
1942 Vegan (Asimov 154)

By tracing the arrival of such science fictional words in the English language we are able to track the humanity's expanding notion of its place in the universe. Significantly all of these[14] are post-Gernsbackian and only arose within the genre of science fiction which fostered continual speculation about the application of scientific knowledge.

Part II: Metamorphosis

Metamorphosis of one kind or another is a common and complex theme in science fiction, with a number of different overlapping and interconnecting strands.

Metamorphosis is frequently the product of mutation, generally as a direct result of exposure to radiation, as from a nuclear accident or the like. This type of metamorphosis is often portrayed negatively, especially in post-apocalyptic science fiction, as in the fear-inspired dictum "Accursed is the Mutant" in John Wyndham's 1955 classic *The Chrysalids*. On the other hand, mutation is also frequently seen as beneficial to humanity, giving rise to such superhuman powers as telepathy, telekinesis, super-intelligence, and ESP. At other times the phenomenon is characterised ambivalently, for example in the comic book character The Hulk, and in the numerous Godzilla movies, where the mutated person/animal whilst creating much destruction and havoc in the end behaves benevolently and beneficially towards humans. An early seminal text in the genetic mutation subgenre is Stanley G. Weinbaum's *The Adaptive Ultimate*, 1935, in which the earliest science fictional occurrence of the noun *mutant* is to be found,[15] after which it becomes a popular motif.

[14] Ignoring the exceptionally isolated occurrence of the word *Sirian* in Voltaire's short fantasy – not science fiction story – "Micromegas" (appearing in English first in Tobias Smollett's 1762 translation (254), and thence reoccurring in subsequent translations, such as the anonymous 1840 translation in the New York *Knickerbocker* magazine (129)).

[15] *Mutant* had been part of scientific vocabulary since 1901, especially with reference to genetic research and experimentation on bacteria, fruit flies and the

An alternate path to metamorphosis is through technology, with which it becomes possible to alter living creatures, including humans. The history of this theme in science fiction can be dated back at least as far as H.G. Wells' *The Food of the Gods and How It Came to Earth*, 1903, where scientists create a chemical foodstuff that has the ability to accelerate growth and create gigantic lifeforms. Subsequent genre science fiction deals with a host of possibilities. While the concept of genetic engineering has only entered public consciousness in the last few decades, the term *genetic engineer* first appeared in science fiction (as opposed to real science) as far back as 1954 (Anderson qtd in Sheidlower). Keeping one step ahead of the times, science fiction writers have created the portmanteau words *gengineer* for 'genetic engineer' (Easton 153) and *genhanced* for 'genetically enhanced' (Hudson 23). These are the result of a common science fiction literary device in which texts are liberally sprinkled with new morphological constructions (the meaning of which are generally transparent to the reader) which are calculated to suggest the popular familiarity of the referent in the fictionalised future. The 1960s saw the introduction of the word *bionics* (Silverberg, "Dark" 27) to refer to human beings augmented by technological devices. By 1973 we had the *personality reconstruction* (Silverberg, "In The Group" 28) and 1982 ushered in the idea of connecting mind and computer directly via *cyberspace* (Gibson 72), the last now appropriated by standard English to refer to the Internet. Will the future see *gengineer* and *genhanced* also become part of standard English?

Finally, metamorphosis is also an outcome of evolution, which in science fiction is concerned with future evolution, as opposed to real science's interest in the past. Influential science fiction editor John W. Campbell, helming *Astounding Science Fiction* from 1937 until his death in 1971, is generally praised for bringing hard science to science fiction and for encouraging stories where "the scientifically minded man" used "reason and practical know-how to solve seemingly insurmountable problems" (James and Mendelsohn 38). However, at the same time, Campbell considered "telepathy and other forms of extrasensory perception [just as] valid scientifically as any speculation about alien environments or rocket ship engineering" (40) and held that these abilities are essentially incipient in the human species, just waiting for the right impetus to cause them to evolve to their full potential. A superb example

---

like. The science fictional use refers to human mutation, given a separate and distinct definition in the *OED*, with earliest evidence from 1938, slightly later than that given here (*OED Online*). Another early example comes from and H.G. Wells' late piece *Star-Begotten*, 1937.

of this motif is to be found in Philip E. High's *Come, Hunt an Earthman*, 1973, in which humanity, when on the backfoot from invading galactic baddies, spontaneously develops worldwide telepathy in order to pool analytical resources and develop super-weapons which ensure eventual victory. Science fiction historian Michael Ashley notes that by the 1950s "stories of psychic and transcendental power began to dominate *Astounding*" (260) and this had a direct influence on science fiction in general. Thus in the 1950s we see the arrival of such terms as *precog* (Dick 10), a person with pre-cognition, that is, the natural ability to see or know the future; *psionics* (Williamson 1951) a more scientific-sounding term for psychic abilities;[16] and *empath* ('McIntosh' 30) for a person with empathic abilities.

Whether via mutation, technology or evolution, much science fiction deals with the metamorphosis of humanity, often to the utmost extremes of possibility or imagination. In 1935 Olaf Stapledon envisioned humans transforming into a superior species he named *Homo superior*, a term picked up by many subsequent writers (Harness 151; Brunner 87; 'Muller' 123; Kapp 102). Indeed, depictions of future humanity are by and large complimentary. For example, although H.G. Wells' evolution of humanity into the meek and blissful but apathetic Eloi, and the cannibalistic, subterranean Morlocks is well-known, this unflattering division into Jekyll and Hyde types did not become a staple of subsequent science fiction. However, another more flattering Wellsian figure, inspired by Darwinism, has become a well-accepted science fiction trope: namely that of future humans with bulbous heads (to hold enormously developed brains) and weedy bodies with vestigial limbs (a result of the absence of physical labour over an evolutionary time scale). This metamorphosed human first appears in one of Wells' earliest short stories, *The Chronic Argonauts*, 1888:

> He was a small-bodied, sallow faced little man … the bones and muscles of his face were rendered excessively prominent and distinct by his extreme leanness. The same cause contributed to the shrunken appearance of the large eager-looking grey eyes, that gazed forth from under his phenomenally wide and high forehead. It was this latter feature that most powerfully attracted the attention of an observer. It seemed to be great beyond all preconceived ratio to the rest of his countenance. (313)

[16] Derived from the Greek letter *psi* (*Ψ*), used to refer to paranormal *psychic* phenomena, and the ending of the word *electronic*.

The same figure appeared again in his essay *The Man of the Year Million*, 1893. The notion was also explicated by Hugo Gernsback:

> Thus a man whose grandfather and father were hard-working laborers, will nine times out of ten inherit a bony as well as robust body and a relatively small brain. Whereas a man whose grandfather and father were mathematicians nine times out of ten will have a comparatively small boned body, which is far from robust, but his brain will be large. (1915, 314)

This, originally science fictional, character eventually gave rise to the ufological *grey*, a member of a group of super-technological Earth-visiting aliens who made their screen debut in Steven Spielberg's 1977 film *Close Encounters of the Third Kind.*

Part III: Beyond Infinity

After humans have conquered the entirety of interplanetary, interstellar and intergalactic space and metamorphosed into more evolved or otherwise superior hominids or superhumans, where might they go from there? What type of being is possible, what type of new, uncharted place? Since at least the 1930s science fiction has had the answer: humans will continue metamorphosing, evolving, refining, each step exceeding the last in increased powers and abilities, eventually dispensing with the physical body altogether and becoming beings of *pure thought* (Peterson 33; Gillings 5); beings who have no physical body but exist as some energy or force that can control matter, space and time at will. This ultimate god-like state is, at least according to Campbellian science fiction, the inevitable end point of unstoppable human progress. And, after humans have spread from their homeplanet Earth and colonised the Solar System, and then outwards through the enumerate star systems of the Milky Way galaxy, and intergalactically throughout the entire Universe, and then transdimensionally through all dimensions of time and space, and there are no more frontiers to be conquered, will they sit Alexander-like and cry? Science fiction says No: beyond the furthest reaches of physical space there is *Infinity*! Well parodied in the motto of Toy Story character Buzz Lightyear – 'To Infinity and beyond!' – the notion of reaching or exceeding infinity has been a staple of much B-grade genre science fiction. *The Internet Speculative Fiction Database* records no less than 352 short story and novel titles containing the word *infinity*, including three separate short stories specifically entitled *Beyond Infinity*. Other titles include: *Arc of Infinity* / *The Infinity Machine* / *Crown of Infinity* / *Escape to Infinity* / *A*

Trip to Infinity / Children of Infinity / Meet Me at Infinity / Split Infinity / Black Infinity / This Side of Infinity and *The Infinity Plague*. A typical example of the way in which infinity is portrayed in these texts is provided by the pen of the prolific Z-grade science fiction writer R.L. Fanthorpe, writing under the Badger Books house name "Karl Zeigfreid": "Mike Sterne reappeared through the rectangular metal gateway of the X-dimensions. He had escaped to infinity, and he had come back" (147). Fabulous nonsense, to be sure. At no time in the novel is the nature of this infinity that the hero escapes from, let alone what lies beyond it, ever elucidated. The very phrase "beyond infinity" is, of course, a plain and simple contradiction in terms as there is no end to infinity, and so there can be no beyond it.[17] In one way this science fictional *infinity* is a place that represents the ultimate pinnacle of human achievement or progression, where humans take on omnipotency and immortality, transcending not only merely physical existence but also the laws of physics and reality. On another level, the use of *infinity* brings home the very finite dimensions of the English language. Beyond all places in the known universe, beyond all boundaries of dimension and time, English reaches its upper limit of hyperbolic description with the word *infinity*, and to go beyond it moves into the world of the ridiculous. The word stock is exhausted at this point and any cogent conceptualisation of something more is brought to a full stop. And yet, while the verbal journey may have finally run into a dead end, science fiction is still moving ever forwards, pursuing the outward urge, undergoing its own metamorphoses, and ensuring that the *Historical Dictionary of Science Fiction* will remain an ongoing endeavour.

Works Cited

Anderson, Poul. "Kyrie." *Black Holes*. Ed. Jerry Pournelle. 1968. London: Futura Publications, 1978.

Ashley, Michael. *Transformations: The Story of the Science-Fiction Magazines from 1950 to 1970*. Liverpool: Liverpool University Press, 2000.

Asimov, Isaac. "Homo Sol." *The Early Asimov: 2*. 1941. St Albans: Panther, 1974.

—. "The Hazing." *The Early Asimov: 2*. 1942. St Albans: Panther, 1974.

[17] This is not unlike the science fiction use of the term *absolute zero*. Because current scientific theory holds that absolute zero or zero Kelvin is impossible to reach (or at least impossible to measure if reached) a number of science fiction stories revolve around technologically attaining this unattainable goal (Anderson; Broxon; Bear).

Astor, John Jacob. *A Journey in Other Worlds: A Romance of the Future*. New York: D. Appleton, 1894.

Banks, Michael A. "Hugo Gernsback: The man who invented the future" *Society for Amateur Scientists*. 14 Oct. 2007. <http://www.sas.org/tcs/weeklyIssues/2004-09-03/feature1/>.

Bear, Greg. *Heads*. 1990. London: Arrow Books, 1991

Behn, Aphra. *A Discovery of New Worlds | From the French | Made English by Mrs. Behn*. London. 1688.

'Binder, Eando'. "Blue Beam of Pestilence." *Amazing Stories* Dec. 1937.

Blair, Andrew. *Annals of the Twenty-ninth Century, or, The Autobiography of the Tenth President of the World-Republic*. London: Samuel Tinsley, 1874.

Brewster, David, Sir. *More Worlds Than One: The Creed of the Philosopher and the Hope of the Christian*. New York: R. Carter, 1854.

The British Apollo, or, Curious amusements for the ingenious. To which are added the most material occurrences foreign and domestic. Perform'd by a Society of Gentlemen 13 (1708): 2.

Brooman, R.A., ed. *The Mechanics Magazine* 18 November 1854.

Broxon, Mildred Downey. "Singularity." *Black Holes*. Ed. Jerry Pournelle. 1974. London: Futura Publications, 1978.

Brunner, K. Houston. "Fiery Pillar." *New Worlds Science Fiction* 38 1955.

Campbell, John W. *Invaders from the Infinite*. 1932. New York: Ace Books, 1961.

—. *The Ultimate Weapon*. 1936. New York: Ace Books, 1966.

Coblentz, Stanton H. "Into Plutonian Depths" *Wonder Stories Quarterly* Spring 1931.

Dick, P. K. "A World of Talent." *Galaxy Science Fiction* Oct. 1954.

Easton, Tom. "The Reference Library." *Analog Science Fiction – Science Fact* November 1988.

A Fantastical Excursion into the Planets. London: Saunders and Otley, 1839.

Fanthorpe, R.L. *Escape to Infinity*. London: Badger Books, 1962.

Fontenelle, Bernard le Bovier de. *Conversations on the Plurality of Worlds. A new translation with great additions. By a gentleman of the Inner-Temple.* London: R. Withy, 1760.

Gernsback, Hugo. "Baron Münchhausen's New Scientific Adventures." *The Electrical Experimenter* Nov. 1915: 312–314.

—. "Baron Münchhausen's New Scientific Adventures." *The Electrical Experimenter* Jan. 1916: 472–475.

—. *Ralph 124C 41+: A Romance of the Year 2660.* Boston: The Stratford Co., 1925.

—. "Discussions." *Amazing Stories* Jan 1927: 974.

—. ed. *Modern Electrics*. New York: Experimenter Publishing. 1908–1913.

—. ed. *The Electrical Experimenter*. New York: Experimenter Publishing. 1913–1920.

Georgiadou, Aristoula and David H.J. Larmour. "Lucian's Science Fiction Novel True Histories: Interpretation and Commentary" *Mnemosyne: Bibliotheca Classica Batava Supplementum* 1998.

Gibson, William. "Burning Chrome." *Omni* July 1982.

Gillings, Walter H. "The Future of Man." *Tales of Wonder and Super-Science*. Winter 1941.

Greg, Percy. *Across the Zodiac: The Story of a Wrecked Record.* 1880. Westport: Hyperion Press, 1974.

Godwin, Bishop Francis. "The Man in the Moone." 1638. Pizor and Comp 9–46.

Harris, John. *Astronomical Dialogues Between a Gentleman and a Lady: wherein the Doctrine of the Sphere, uses of the Globes, and the elements of Astronomy and Geography are explained.* London: John Horsfield, 1725.

Harness, Charles L. *The Paradox Men.* 1953. London: New English Library, 1974.

High, Philip E. *Come, Hunt an Earthman.* 1973. London: Arrow Books, 1987.

Howell, James. *Epistolæ Ho-Elianæ: Familiar Letters Domestic and Forren*. London: H. Moseley, 1645.

Hudson, Edward S. *Alien Death Fleet.* New York: Pageant Books, 1989.

The Internet Speculative Fiction Database. 14 Oct. 2007. <http://www.isfdb.org>

James, Edward and Farah Mendlesohn. *The Cambridge Companion to Science Fiction*. Cambridge: Cambridge University Press, 2003.

Kapp, Colin. *Cageworld: The Tyrant of Hades*. London: New English Library, 1982.

Keill, John. *An Introduction to the True Astronomy: or, Astronomical Lectures, Read in the Astronomical School of the University of Oxford.* London: Henry Lintot, 1739.

Kress, Nancy. "Women in American Science Fiction." 14 Oct. 2007. <http//:www.lysator.liu.se/lsff/mb-r21/Speech_by_Nancy_Kress.html>

Larbalestier, Justine. *The Battle of the Sexes in Science Fiction.* Middletown. CT: Wesleyan University Press, 2002.

Leinster, Murray. "Proxima Centauri." *Before the Golden Age III*. London: Futura, 1975.

'McIntosh, J.T.' "Empath." *New Worlds Science Fiction* Aug.1956.

'Muller, John E.' *Night of the Big Fire*. London: Badger Books, 1962.

"mutant, *n.* and *a.*" *The Oxford English Dictionary*. June 2003. *OED Online*. Oxford University Press. 14 Oct. 2007. <http://dictionary.oed.com/cgi/entry/00319638>.

Olsen, Bob. "Four Dimensional Transit." *Amazing Stories Quarterly* Fall 1928.

Peterson, John Victor. "The Reinmuth Rider." *Thrilling Wonder Stories* June 1938.

Pizor, Faith K. and T. Allan Comp. Eds. *The Man in the Moone: An Anthology of Antique Science Fiction*. London: Sidgwick and Jackson, 1971.

Proctor, Richard A. *Other Words Than Ours: The Plurality of Worlds Studied under the Light of Recent Scientific Researches*. London: Longmans Green And Co., 1870.

—. *Other Words Than Ours: The Plurality of Worlds Studied under the Light of Recent Scientific Researches*. 4th ed. New York: A.L. Burt, 1878.

Prucher, Jeff, ed. *Brave New Words: The Oxford Dictionary of Science Fiction*. New York: Oxford, 2007.

Schachner, Nat. "The Son of Redmask." *Astounding Stories* August 1935.

Sheidlower, Jesse. *Science Fiction Citations*. 14 Oct. 2007. <http://www.jessesword.com/sf/>

Silverberg, Robert. "Dark Companion." *Amazing Stories* January 1961.

—. "In The Group." *The Shape of Sex to Come*. Ed. Douglas Hill. London: Pan Books, 1978.

Smith, Clarke Ashton. "Marooned in Andromeda." *Other Dimensions I*. 1930. St Albans: Panther, 1977.

—. "The Monster of the Prophecy." *Out of Space and Time II*. 1931. St Albans: Panther, 1974.

Smollett, Tobias, T. Francklin et. al. "Micromegas: A Comic Romance." *The Works of M. de Voltaire. Translated from the French*. Vol. 11. London: J. Newberry, 1762: 250–282.

"space, *n*[1]." *The Oxford English Dictionary*. 2nd ed. 1989. *OED Online*. Oxford University Press. 14 Oct. 2007. <http://dictionary.oed.com/cgi/entry/50231942>.

Stapledon, Olaf. "Odd John." 1935. In Prucher, Jeff. *Brave New Words: The Oxford Dictionary of Science Fiction*. New York: Oxford, 2007.

Stapleford, Brian. *Historical Dictionary of Science Fiction Literature*. Lanham, Maryland: Scarecrow Press, 2004.

Sterling, Bruce. *Schismatrix*. Harmondsworth: Penguin Books, 1985.

Voltaire. "Micromegas." *The Knickerbocker: Or, New Monthly Magazine*. Aug. 1840: 126–143.

Wallace, Alfred Russel. *Man's Place in the Universe: A Study of the Results of Scientific Research in Relation to the Unity or Plurality of Worlds*. London: Chapman and Hall, 1904.

Wells, H. G. "The Chronic Argonauts." *Science School Journal*. Apr. South Kensington: Normal School of Science and Royal School of Mines, 1888.

—. "The Man of the Year Million." *The Pall Mall Gazette*. London. 9 Nov. 1893.

—. "The Food of the Gods and How It Came to Earth." *Pearson's Magazine*. London: C.A. Pearson. Dec. 1903.

—. *Star-Begotten*. 1937. London: Sphere Books, 1975.

Whewell, William. *The Plurality of Worlds*. Boston: Gould and Lincoln, 1854.

Weinbaum, Stanley G. "The Adaptive Ultimate." *A Martian Odyssey and other Classics of Science Fiction*. New York: Lancer Books, 1962.

Williamson, Jack. "Collision Orbit." *Astounding Science Fiction* July 1942.

—. "The Greatest Invention." *Astounding Science Fiction* July 1951.

Wyndham, John. *The Chrysalids*. London: Michael Joseph, 1955.

PLAYING THE WORLD: POSTCOLONIAL COMPUTER GAMES AS SPATIAL STORIES

SYBILLE LAMMES

Abstract

Simulation games like *Age of Empire*, *Victoria* and *Civilization* quickly spring to mind as prime examples when imperialism and colonisation in games is being discussed. This comes as no surprise since the player of such games is both a traveller through the world of the game as strongly involved in actively, colonising, manipulating and transforming this world. To a greater extent than in many other games she is explicitly invited to master landscapes, and to form a strong and lasting nation. Hence such games pose interesting theoretical questions about which functions imperialism and colonialism may entail when space is "in the hands" of the player.

In this article I investigate which cultural functions of imperialism and colonialism can be recognized in games that invite players to conquer, shape and map new territories. In these (what I will call postcolonial) games the player undertakes a voyage that hinges on notions of a colonialist militaristic exploration, turning the gamer at once into a new world traveller, ruler, military force and explorer who surveys and masters and creates domains. I will examine how such games can be understood as contemporary expressions of the aftermaths of colonisation and imperialism. More specifically I wish to scrutinize whether and how the interactivity of these games gives the player more opportunities to be part of the politics of representation that are part of the aforementioned.

Introduction

There appears to exist no academic contestation over the claim that spatiality is an essential quality of all computer games. As has been convincingly asserted before, the player of a game always seems to move through a spatial environment where she tries to find her way and by doing so proceeds to score points and rise to higher levels. Hence all computer games are fundamentally about space (Aarseth 2000; Aarseth 2004; Newman). However, a more contested notion when it comes to computer games is that of narration. Since game-studies started to name itself as a discipline in its own right, a sometimes heated and somewhat over-dramatized debate took place in which some so-called ludologists denounced the notion of narrative as being a central characteristic of games and others maintained that games were stories in essence, too.[1]

This article is about a specific kind of game in which both spatiality and narration are important organizing principles. The games that I wish to discuss, such as *Age of Empire*, *Victoria*, *Rise of Nations*, and *Civilization*, invite the player on a journey where she encounters landscapes that have to be claimed by means of militaristic fortification, scientific developments and economic growth. This explorative journey is set in a historical sphere and the player becomes a ruler of the past, fortifying her nation by appropriating and colonising landscapes. Clearly spatial exploration and the development of a player's (his)story are part and parcel of such games.

In this article I will approach such games as postcolonial expressions that enable the player to handle the mastering of space in a particular contemporary way. The player is invited to become more than a traveller, but also an interactive explorer who can, to a certain extent (that is: within the parameters of the games), change the way that borders have been drawn. In that sense the player has more in her hands than just frozen ideological representations of colonial space. It is precisely the offered ability to transform spatial formations that endows these games with such

---

[1] To name but a few that participated in this debate: G. Frasca, "Ludologists Love Stories, Too: Notes from a Debate That Never Took Place," *Proceedings of Level Up* (2003), Gonzalo Frasca, "Ludology Meets Narratology: Similitude and Differences between (Video) Games and Narrative," *Ludology* (2005), vol. 2008, Jesper Juul, "The Definitive History of Games and Stories, Ludology and Narratology," *The Ludologist* (2004), J. Murray, "The Last Word on Ludology V Narratology in Game Studies." *Proceedings of DiGRA 2005 Conference: Changing Views of Worlds in Play* (2005), J. Simons, "Narrative, Games, and Theory," *Game Studies* 7.1 (2006).

strong postcolonial qualities: these games are not only dealing with the "reproduction" of colonialism in the present, but are also concerned with transformations of colonial representations and practices.

To specify how this particular process of transformation takes shape, I will approach such games as spatial stories as defined by de Certeau. Since the interplay between frozen spatial perceptions and more fluid spatial experiences play such a central role in de Certeau's perception of spatial stories, I will use his theory to show how such games entail specific spatial transformations that can be called postcolonial.

Post-colonialism

That the games that I wish to discuss in this article have a colonial dimension, doesn't need much explanation: the player is invited to conquer and master environments and that endeavour is clearly defined in colonial terms. You choose a historical leader such as Elizabeth I and are expected to win territory through exploring land, finding resources, founding fortresses, creating alliances and enemies, strengthening your military apparatus, etc. But why are they post-colonial means of expression? I would argue that they should not just be called this for the simple reason that they are produced in a post-colonial era for post-colonial users. These games do much more than re-enact the past but rework this past, by handling space (and time) as changeable and anachronistic. They are therefore a contemporary and postcolonial means of expression that revisit and where colonial history is remoulded in order to make sense of a present through spatial exploration (Lammes; Magnet).

In the humanities post-colonialism is often used as a term to describe the repercussions and aftermath of a long western history of dominance and exploration of other lands and cultures. It refers to how the echoes and consequences of this colonial mentality are still palpable, be it in (diasporic) life stories, literature, films, or social geography. Or as social geographer Johnson defines post-colonialism: "A critical politico-intellectual formation that is centrally concerned with the impact of colonialism and its contestation on the cultures of both colonizing and colonized peoples in the past, and the reproduction and transformation of colonial relations, representations and practices in the present" (Johnston).

In my opinion, it is this last feature of *transformation* that is most interesting when thinking about these digital games in relation to post-colonialism. Due to their interactive character, they offer players many opportunities to remake and reshape their colonial pasts through spatial exploration. They offer openings to make playful connections between spaces that may not be historically accurate (e.g. making winners out of colonial losers) but that are exactly about experiencing the consequences of being a colonizer. This can be related to what I have called elsewhere "a postcolonial and contemporary disorientation of belonging" in which spatial "history cannot be easily retold in a singular way" (Lammes 129). To come back to the cultural spatial functions such games may serve, they offer possibilities to reshape and make sense of ideologies of appropriation and colonial expansion whose consequences are still so much intertwined with our contemporary daily life.

Spatial regimes

So the games that I am discussing in this article can be best called post-colonial because transforming colonial space is such a key activity for any player. But how exactly do these games deal with this transformation of space in an effort to make sense of the post/colonial world of which they are part and parcel? Which spatial strategies are available to the player to play with the rules and consequences of our colonial past, i.e. to transform colonial spaces?

In this article I will try to answer this question by applying de Certeau's notions of spatial stories to such games. The reason that I have chosen for this perspective is that the interplay and friction between ideological spatial rules and subjective personal appropriation of space is so central to de Certeau´s argument. It offers therefore an interesting theory to understand how the player can interactively deal with colonial spaces and can make them her own through transformation.

I will look into four main concepts of de Certeau that have been circulating in game-studies so far, namely space, place, the tour and the map. Place and maps are perceived as abstract, timeless and stable. It is on this level that frozen ideological representation of colonial space should be situated. Space and tours on the other hand are seen as concrete, known and changeable and part of a more personal experience. It is on this level that player can interactively build her own notion of empire and space and the aspect of transformation that is so central to post-colonialism is being fore-grounded.

Since these terms are often referred to in a rather condensed way I will start by re-visiting de Certeau's writings on these matters before discussing how they have been used in relation to games until now. Furthermore, I will argue that de Certeau's understanding of boundaries should be more explicitly added to this quartet to come to a better understanding of the studied games.

Stories as playgrounds

Since game-studies has announced itself as a discipline, de Certeau's notion of spatial stories has been applied to computer games. It was especially a co-written article or more precisely dialogue by cultural theorist Henry Jenkins and literary historian Mary Fuller, "Nintendo® and New World Travel Writing: A Dialogue," that set the trend. In this article it is asserted that Nintendo games, and even new media as a whole, may be best perceived as spatial stories. Jenkins observes that games and other digital expressions can be best described as spatial narratives since the player primarily constructs a narrative by travelling through space (Fuller and Jenkins 1995). This idea was soon taken on board by Friedman in his writing about *Civilization*. He takes Jenkins' remark a step further by claiming that it is foremost a suitable concept for understanding simulation games in which the player re-enacts and gives shape to history through the mastery of landscape.

Jenkins' and Friedman's views have since then been the touchstone for many academic writings on games to follow (Magnet 2006; Poblocki 2002; Newman 2004; Lammes 2003). Following Jenkins and Friedmann, these authors mostly refer to the difference that de Certeau made between space and place and between the map and the tour and try to apply these concepts to the aforementioned games.

De Certeau re-visited

In *The Practice of Everyday Life* de Certeau dedicates a chapter to the working of spatial stories. Unlike some ludologist game researchers would claim, spatial stories are everywhere according to de Certeau. Even more than that, spatial stories are the main way in which humanity makes sense of everyday life: They "take us places." Spatial stories are the main organizing principle for all human activity and are especially important when it comes to trying to come to grips with spatial change: "[E]very day, they traverse and organize places; they select and link them together, the make sentences and itineraries out of them. They are spatial trajectories" (de Certeau 115, 1984).

Correspondingly, spatial stories are part of our strategies to make sense of our everyday life according to de Certeau. And as such they are everywhere and ubiquitous. To understand how they can be understood he makes the aforementioned difference between space and place and map and tour. In the following paragraphs I will explain these terms in more detail.

Space/Place

The most important difference between place ("lieu") and space ("espace") is that the former is about stability and an ordered configuration of elements whilst the latter rather implies mobility and has a "polyvalent" character. Place refers to the "proper" order, to the way spatial positions are related in an objective account, whilst space is about how we deal with such places, how we live them as "a practiced place" (de Certeau 1984, 115-16). De Certeau gives the example of walking the streets to explain what he means by this. The configuration of the streets he equates with place, while the act of travelling through these streets changes them into space: "Thus the streets geometrically defined by urban planning is transformed into a space by walkers" (117). So, as place is set and univocal, the notion of space has as manifold meanings as there are walkers.

De Certeau speaks of both terms as "determinations in stories" that constantly influence each other, changing "places into spaces and spaces into places" (118). He identifies place as a "determination through object" (ibid.), as the purpose to make static and lifeless objects. Space, on the other hand, presupposes a subjective purpose. It implies movement and change and can be associated "with a history" (ibid.). In stories, these two determinations should be understood as "passages back and forth" (ibid.) in which a lifeless objective place can become an animated and changeable space and space can likewise be frozen into a place.

Map/Tour

De Certeau introduces the difference between the map and the tour as "analytical categories" (118) to distinguish the different modes of this interplay in one of the most basic travel stories, namely that of a spatial description (de Certeau 1984, 119). From a former study of how New York residents describe their apartments, he learns that an overwhelming amount of participants described their dwellings in terms of *going*, and that only a small minority used terms of *seeing* to explain how their apartments looked. The telling mode of *seeing* he links to the notion of the map. A map can then be described as a static representation of the world we live in. It objectifies spatial relations. The *going* mode he links to the notion of the tour. A tour, a dynamic principle that is subjective since the point of view of the traveller is central. According to de Certeau, these two conceptions of spatiality are both incongruous dimensions of contemporary culture: we are confronted with static ("scientific") representation of the world we live in, while at the same time sensing our space in a dynamic and more personal way. As place and space, maps and tours necessitate one another and come into being through a two-way-movement. Even more so, a map always presupposes a tour, since one needs to go somewhere to give an objective spatial account of it (119-20).

De Certeau argues that although these terms are in essence intertwined, maps have become more and more separated of the experience of a tour in western cultures. He states that their interaction is still very clear on medieval maps where the itineraries themselves remain basic while the emphasis lies on how long it would take to travel and where to stop, hence a map being "a memorandum prescribing action" (120). Gradually maps erased the traces of the touring that permitted their making. Maps became more formal and abstract and "[t]he tour describers have disappeared" (121). De Certeau states that in contemporary western culture we have to deal with this isolated "system of geographical places" (ibid.) simultaneously experiencing space as on ongoing and mobile practice. Everyday stories are according to de Certeau ways to make sense of such frictions: "they are treatments of space" (121).

New world dialogues

In the aforementioned dialogue between Jenkins and Fuller, Jenkins uses the writing of de Certeau to situate Nintendo games in a wider landscape of popular entertainment. Additionally, his dialogue with Fuller is aimed at comparing the treatment of space and place in Nintendo games with earlier Anglo-Saxon writings on the discovery of new lands. Consequently an interesting picture emerges of the imagination of space as a popular discourse, one that is closely related to the American myth of the Wild West and new frontiers.

He finds that de Certeau's ideas on spatial stories are very apt for explaining such imaginary quests as stories. According to Jenkins Nintendo games are spatial stories because they cannot be easily defined in terms of plot or character. The story mainly develops by moving through landscapes whereas the plot remains relatively simple or static (Fuller and Jenkins).

In this respect Jenkins places Nintendo games in the same league as early SF pulp fiction and imaginary quest like *Alice in Wonderland* where the story also develops through an exploration of unknown territory and plot and character are of a minor importance. He also sees a parallel with playground attractions such as the rollercoaster. Like such attractions Nintendo games offer the recreational sensation of sight-seeing, while the landscape does not seem to hold any economical value. But contrary to this attraction they offer the participant the opportunity to interact and transform the landscape during her tour.

When applying de Certeau's distinction between place and space to Nintendo games, Jenkins relates place to the code that presupposes the game, while space "happens" at the moment that the code is in the hands of the player who creates her own imaginary landscape that is personal and mobile. The difference of the map and the tour is conceived more literary by Jenkins, when he ascertains that the maps that are supplied with the game (on paper) are more like the medieval maps of which de Certeau speaks, in which descriptions are less isolated from the tour that they facilitate by giving more description on how to move to the game than static representations.

The Map as Space: Friedman and his Discontent

Friedman believes that simulation games like *Civilization* differ from the games that Jenkins describes in the way maps are used by the player. Dissimilar to what Jenkins claims, such games are more about inviting the player to become a mapmaker, so he argues: during the playing of the game one identifies with the landscape (a spatial practice) in a more abstract sense because one constantly sees the landscape from above (Friedman).

Because simulation games fix the player in a depersonalized frame of mind, they can tell their story in the abstract, without ever bringing it to the level of individual experience. The map is not merely the environment for the story; it's the hero of the story. Thus, because the environment is the main focus of such games Friedman asserts that the player has less of a subjective identification with the visited landscape and that space is experienced as mapping.

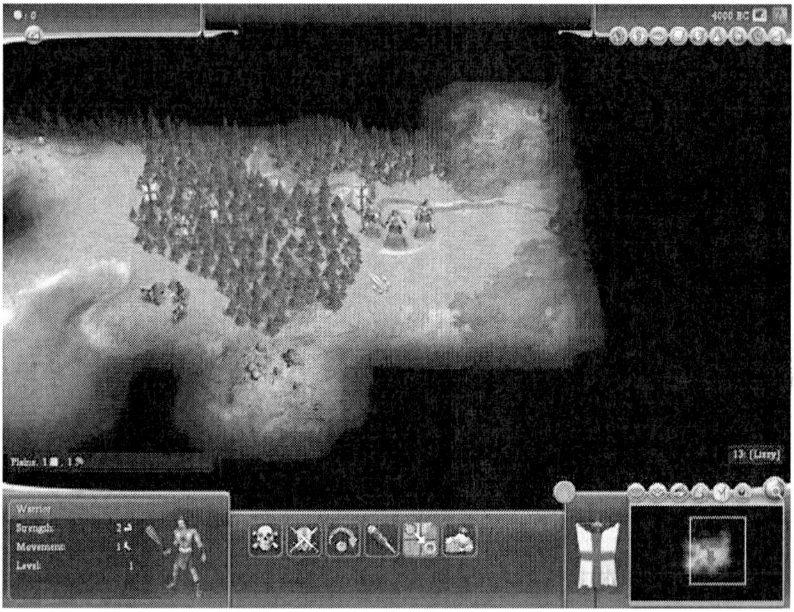

Screenshot from Sid Meier's *Civilization IV* (Firaxis Games, 2005)

At this point Friedman mixes up his spatial metaphors. He uses the term space from the distinction space/place that de Certeau uses as two determinations in *any* story and the term map from the map/tour distinction which de Certeau employs to analyze the working of descriptions of voyages. Also, he misses the important point of de Certeau that stories can oscillate between space and place and map and tour.

How important this interplay is in such games, I would like to illustrate with the game *Age of Empires* (AOE). In AOE the player is in a constant flux of moving through territory, which is translated into an expansion (filling in) of the mini map. Conversely, I can click on the mini map to move to an area on the big screen. I can even click on my explorer, go back to the mini-map, click on the area I want to send him to and he will go to that spot. Hence mapping and touring entertain a highly dynamics relationship in this games, which somewhat undermines Friedman's claim that the player simply becomes a cartographer as opposed to a tour-maker. She indeed becomes a mapmaker, but this cannot be described as a depersonalized endeavour. It would be more precise to call the player a cartographer on tour.

Screenshot from *Age of Empires III* (Ensemble Studios, 2005)

Furthermore, when he states that the map becomes the protagonist of the story with whom the player identifies, he overlooks a pivotal quality of any game: it is always the *player* who becomes the hero (or the loser). Games are no pre-set texts with pre-set heroes, but are interactive forms of expression. The player has an enormous influence on the outcome of a game and thus a different identification process emerges in which a map cannot be a protagonist, but the player dynamically fluctuates between the position of being a cartographer and traveller at once. To bring this back to the central argument of this paper: such games offer the player more than a frozen ideological representation of empire and colonial space as maps have become, but allow her to spatially transform her environment, hence making sense of it in her own way. It is at this level that spatial ideologies become messier and paradoxical qualities of postcolonial cultures seep through.

What Friedman probably means to say is that identification is such games takes even more place on a spatial level than the Nintendo games, since you don't play with an avatar and are constantly involved in mapping (getting an overview) of the space that you create through touring mostly from above instead of on virtual foot. Hence a strong interplay can emerge between mapping and touring and the gamer is constantly and explicitly involved in making her own space, translating that in places and vice versa. As such these stories liberate us from the isolated, unchangeable and static form maps have become after the renaissance. They bring maps back into life. Or to put it in terms of identification: the player becomes her landscape.

Like Friedman, I do think that such computer games are indeed the pinnacle of what de Certeau defines at spatial stories, precisely because spatial exploration is so central to them and is not hindered by plot or avatar (character). But not for the reason that they give the player the opportunity of "experiencing space as a map," but because of the constant interplay they allow her between the static and the mobile, the subjective and the objective.

An important difference between the map and the tour that is somehow missing from Jenkins' and Friedman's account can clarify this further. As I stated before de Certeau links mapping to the act of *looking* and *going* to the act of touring (118). In the case of AOE looking and going are interwoven processes and a description of playing the game is a constant oscillation of seeing and going.

In the example I just mentioned, looking and exploring are constantly feeding into each other. The mini map that you can see at the left corner of the game may at first glance look like a static entity; as something to look at, rather than to manipulate. Yet, it is a map that I can change. Firstly, it becomes filled in by every movement that I let my explorer make, an act of going. While the black recedes on the main screen, the map becomes more filled in as well. So the touring of my explorer feeds directly back into the map. Secondly, when I point my cursor at a point in the map, and click my mouse, I am transported to that point on the main screen. Again, an act of going. Hence it is a map with touring qualities. This oscillation between looking and going is a crucial activity when I play the game and again draws attention to the powerful postcolonial functions of such games: maps become more fluid and transformable, hence less framed in dominant ideologies of empire in which maps can play such a central role as means of solidification.

Stories and boundaries

I would argue further that the games that are under scrutiny in this article are not only spatial stories *par excellence* because they foreground the interplay between map and tour. They also accentuate two other figures that are important to de Certeau's definition of spatial story: the frontier and the bridge, which are again important figures when trying to understand what makes such games such powerful postcolonial means of expression.

De Certeau argues that stories perform an important function in everyday life by setting limitations. By describing space they arrange and order cultural domains. As such they do not only set limits but also alter boundaries: "[…] one can see that the primary function is to *authorize* the establishment, displacement or transcendence of limits" (de Certeau 1983, 123). To describe this paradoxical quality of boundaries, he distinguishes two narrative forms in every story that respectively have the power to fix boundaries and to revise them, namely the *frontier* and the *bridge*.

For explaining the figure of the frontier de Certeau takes his reader on a etymological tour, in which he shows how the Roman ritual act of founding (fās) precedes any military or political action. He claims that this tendency of putting frontiers down before the action takes place can still be discerned in contemporary western cultures and the way nationhood is perceived. Be it more scattered, the "primary role of stories" still is that after their formation a space is formed that can function as a playground for actions (ibid., 124). Such actions can nevertheless also transgress the limits that are first set by the story (e.g. feudal conflicts in which set borders are contested). As such boundaries are the prerequisite for any social practice.

When the frontier becomes a point of passage, de Certeau speaks about the figure of the bridge: "It turns the frontier into a crossing, and the river into a bridge. It recounts inversions and displacements: the door that closes is precisely what may be opened; the river is what makes passage possible; the tree is what marks the stages of advance; the picket fence is an ensemble of interstices through which one glances pass" (de Certeau 1984, 128). He states that the frontier as a spatial narrative figure always contains "this logic of ambiguity" (Ibid.). Seen from this perspective every story is about setting frontiers to be able to cross them.

In this process the figure of the frontier and the bridge entertain a paradoxical relationship. In de Certeau's vocabulary the frontier should be equated with the "(legitimate) space" (126), since it creates spaces such as nations and empires. The bridge, on the other hand, points to what is not part of that space, in other words to "its (alien) exteriority", such as other nations, unconquered territories or no-man's lands (Ibid.). But, as he indicates, they entertain a dynamic and paradoxical relationship. The frontier has a mediating or "bridging" quality in that it is at the same time the point of contact between the two entities it separates. In itself it does not belong to either entity, or as he puts it: "The river, wall or tree *makes* the frontier" (Ibid.). Frontiers are in that sense twilight zones.

More than the goalposts and secret passages that Jenkins mentions as points of transgression in Nintendo games, such games explicitly ask the player to mark off land and to both create borders and transgress them in an interactive way. As a player I have to deal with at least two different kinds of borders at once: the border between filled in territory and undiscovered black space and the borders between my own territory and that of others.

Interestingly, games like AOE foreground the bridging qualities of borders. Since the constant transgression is so central in such games' borders have a semi-permeable quality. They are always crossings to "(m)otherlands" and borders are never presented as hard lines.

Screenshot from Sid Meier's *Civilization IV* (Firaxis Games, 2005)

Of course the map that has to be filled in has preset limits, on what Jenkins calls the level of code. And also, demarcating your territory is a central activity. Yet the way your own territory is shown on screen is rather fuzzy. In AOE it is presented by different colours for different nations and in *Civilization IV* by an opaque line that is drawn around your zone of influence. As a player, I am involved in a constant process of defining and expanding my own territory while other players can try to diminish my playroom or push back my borders. Hence frontiers are points of passage, be it to another landscape or another level.

That such games can foreground this bridging quality should again be seen in the light of the interactive quality of games. Because the player is able to draw and transcend boundaries herself and to be interactively involved in making her own space, the emphasis lies on the passing of borders. She becomes the enactor instead of the teller of a spatial story, and tries to expand her story by pushing spatial limits.

Relating this to the central question posed at the beginning of this paper, the interactivity of these games offers the gamer means to play with the ambiguities of nationhood and empire and to play with spatial ideological rules. She can and should employ spatial strategies to transform her empire, thus making sense of postcolonial culture in a more personal way. Instead of just replicating dominant politics of spatial representation she embrases them while at the same time transforming them. As an enactor of a postcolonial spatial story, she becomes empire.

Spatial Functions

It is precisely the tension and oscillation between the abstract and concrete experience of spatiality that is an essential quality of the games that are central to this paper. The player is engaged in touring through and exploring new territory in a highly personal endeavour. During the process of touring unknown abstract place gives way to space that is identified by the changes she makes to the territory. Furthermore, spatial representations such as maps and an overwhelming amount of statistics keep transforming during her journey.

As such, these games are a very important popular domain for expressing and giving shape to the paradoxes of postcolonial culture. The potentials for making the abstract concrete and vice versa, give the player rich grounds for playing with being the empire and the colonizer. It offers her opportunities to make sense of what it means to be a postcolonial being and to experience the transformation of space that it entails first hand.

Works Cited

Aarseth, Espen. "Allegories of Space: The Question of Spatiality in Computer Games." *Cybertext Yearbook* (2000): 152-71.

—. "Genre Trouble: Narrativism and the Art of Simulation." *First Person: New Media as Story, Performance and Game*. Cambridge, Mass.: MIT Press, 2004. 45-47.

de Certeau, Michel. *The Practice of Everyday Life*. Berkeley, London: University of California Press, 1984.

Frasca, G. "Ludologists Love Stories, Too: Notes from a Debate That Never Took Place." *Proceedings of Levelup* (2003).

Frasca, Gonzalo. "Ludology Meets Narratology: Similitude and Differences between (Video) Games and Narrative." *Ludology*, 2005. Vol. 2008.

Friedman, Ted. "*Civilization* and Its Discontents: Simulation, Subjectivity, and Space." New York, 1999. *Discovering Discs: Transforming Space and Genre on CD-ROM*. Ed. Greg Smith. 132-50. New York University Press. <http://www.duke.edu/~tlove/civ.htm>.

Fuller, Mary, and Henry Jenkins. "Nintendo® and New World Travel Writing: A Dialogue." Cybersociety: *Computer-Mediated Communication and Community*. Ed. Steven G. Jones. Thousand Oaks: Sage, 1995. 57-72.

Johnston, R. J. *The Dictionary of Human Geography*. 4th ed. Oxford, UK, Malden, Mass.: Blackwell Publishers, 2000.

Juul, Jesper. "The Definitive History of Games and Stories, Ludology and Narratology." *The Ludologist,* 2004.

Lammes, Sybille. "On the Border: Pleasures of Exploration and Colonial Mastery in *Civilization III Play the World*." *Level Up: Digital Games Research Conference*. Eds. Marinka Copier and Joost Raessens. Utrecht: Utrecht University, 2003. 120-29.

Magnet, S. "Playing at Colonization: Interpreting Imaginary Landscapes in the Video Game Tropico." *Journal of Communication Inquiry* 30.2 (2006): 142-62.

Murray, J. "The Last Word on Ludology V Narratology in Game Studies. *Proceedings of DiGRA 2005 Conference: Changing Views of Worlds in Play* 17, 2005.

Newman, James. "Chapter 7: Videogames, Space and Cyberspace: Exploration, Navigation and Mastery." *Videogames*. London: Routledge, 2004. 107-26.

Poblocki, Kacper. "Becoming-State: The Bio-Cultural Imperialisme of Sid Meier's Civilization." *Focaal, European Journal of Anthropology*.39 (2002): 163-77.

Simons, J. "Narrative, Games, and Theory." *Game Studies* 7.1 (2006).

CONTRIBUTORS

Derya Agiş worked for the Department of Translation and Interpreting of Girne American University in the 2007-2008 academic year. In 2007 she received her M.A. in English Linguistics from Hacettepe University with a thesis entitled "A Comparative Cognitive Pragmatic Approach to the Judeo-Spanish and Turkish Proverbs and Idioms That Express Emotions." She attended a postgraduate course in Translation Studies at the University of Bari in 2002, and received her B.A. in Italian Language and Literature with the highest honors in 1999.

Dimitris Agouridas was born in Athens in 1973. He studied Tourism Management, Tourism Policy, and has also taken courses in Psychology and Social Anthropology. He lectures at Yeditepe University, Istanbul, and is currently pursuing a doctoral degree in Theoretical Psychoanalysis. His research focuses on the articulation of psychoanalysis with tourism.

Mazi Allen is a PhD candidate at the State University of New York in Binghamton. He is currently completing a dissertation, Notions of Resistance: Frantz Fanon's Clinical Studies (1954-1960), detailing the relationship between Frantz Fanon's clinical practice (in Algeria and Tunisia during the 1950s) and the directions taken in his later political thought.

Vassil Hristov Anastassov has more than 20 years of experience in teaching General, Indo-European, Slavic and Balkan Linguistics as well as History of the English Language and English Literature, Semiotics in Literature, Critical Discourse Analysis. Has been involved in research work and in the development of major academic projects (Comparative-Historical Indo-European Linguistics- Bulgarian Academy of Sciences; Language and Culture in the History of Politics). Major interests: Historical Linguistics,Philosphy of Language Participation in various conferences (mainly linguistics and language in politics) in Europe, USA, Canada. Has worked for the last decade in Istanbul at Bilgi University, Fatih University and The Bosphorus University.

Mohamed Bakari was educated at the Universities of Nairobi and York, England, and is currently Chair of the Department of American Culture and Literature at Fatih University, Istanbul. He taught at the University of Nairobi for over two decades before moving to Turkey in 1998. He was a Visiting Fulbright Professor at UCLA 1985/6. He has also been a Visiting Professor at the African Studies Center, University of Indiana, Bloomington, Institüt für Afrikanistik, University of Cologne, University of Bayreuth, Germany, Modern Oriental Institute, Berlin, and was recently African Visiting Fellow at the Oxford Centre of Islamic Studies. Among his major publications is The Morphophonology of the Kenyan Swahili Dialects, published by Detriech Reimer Varlag, Berlin. He has a keen interest in Postcolonial literature, theoretical issues in the structures of African languages, and contemporary issues affecting the future of Africa.

Enrique Banús studied Comparative Literature, Romance and German Philology at the Universities Bonn and Aachen. He has been lecturer at the universities of Paderborn and Aachen in Germany and Navarra in Spain. Cureently he is Dean of the Faculty for Humanities at the Universitat Internacional de Catalunya in Barcelona. In 2003 he was awarded by the European Union with a Jean Monnet Chair ad personam "European Culture." For the term 2009-2010 he is World President of the European Community Studies Association (ECSA).

Emine Yeşim Büyükaslan currently is a Research Assistant in the Department of American Culture and Literature at Fatih University. She completed her B.A. degree at the Department of English Language and Literature at Fatih University in 2002, and started her M.A. in 2005 at the same department and university with her MA thesis entitled “Love in Dystopia.” She got her Professional Tourist Guide Certificate at 2008, and is now a PhD candidate at the Department of Comparative Literature at Fatih University.

Zelma Catalan teaches nineteenth and twentieth century English Literature and Literary Stylistics at the Department of English and American Studies, Sofia University “St. Kliment Ohridski.” Her research interests lie mainly in the field of the Victorian novel. Her publications cover the works of Thackeray, Dickens, Trollope, Charlotte Bronte, Browning, as well as some more recent novelists. She is the author of a full-length study “The Politics of Irony in Thackeray’s Mature Fiction” (forthcoming). Dr. Catalan is Director of the MA Program in Translation at the Department of English and American Studies, Sofia University. At present she is engaged with a major project on Dickens’s fictional worlds.

Ralph Cleminson, Professor of Slavonic Studies at the University of Portsmouth, works on the culture and history of mediæval *Slavia Orthodoxa* with particular reference to the written tradition. A specialist in cyrillic palæography and codicology; he is working, *inter alia,* on the textual transmission of the Slavonic version of the New Testament.

Gül Bahar Cömert-Agouridas is a psychiatrist and psychoanalyst, and a lecturer in Psychiatry at Marmara University, Istanbul. She has worked extensively on issues related to infertility and attachment. Her research interests, however, expand beyond the clinical field, well into the domains of art, the social sciences, and the humanities.

Guita Grin Debert is Full Professor at the Department of Anthropology, Universidade Estadual de Campinas (UNICAMP) Brazil, and a researcher at the Brazilian National Board on Scientific and Technological Development (CNPq). She was Vice-President of the Brazilian Anthropological Association (2002-2004) and editor of the Revista Brasileira de Ciências Sociais (1998-2002). She is author of research, books and articles on violence against women, old age and aging and gender. Currently she is the coordinator of the research center PAGU - Center for Gender Studies at UNICAMP.

Joanna Derdowska, absolvent of sociology and Slavonic literature at Nicolai Copernici University in Torun, is preparing a doctoral thesis about city space in contemporary Czech literature at Charles University in Prague. Her book *Prague metamorphosis. The sacred and the profane in the Prague city space* was published in 2006.

Elżbieta Foeller-Pituch (Center for Historical Studies, Northwestern University) is a literary historian who has published articles on American literature and on aspects of the classical tradition in America, including a chapter on Athena as a cultural icon in the United States in *American Women and Classical Myths*, ed. Gregory Staley (Baylor UP, 2009). She is working on a book-length study of the enduring influence of Greek and Roman myths in American culture, a topic that stems from her research during an American Council of Learned Societies fellowship at Harvard University. Her current research interests also include film adaptations of fiction; the evolution of the mystery genre; and the cultural symbolism of food in literature.

Nagihan Haliloglu was born in Istanbul. She did her undergraduate studies at Bogazici University and studied Oriental Studies at the

University of Oxford. She has recently finished her dissertation on Jean Rhys and representation of female and colonial selves. She has published articles on diasporic writing and the contemporary Turkish novel.

Jeffrey Herlihy is Assistant Professor of Humanities at the University of Puerto Rico at Mayagüez. His work has appeared in *The Hemingway Review*, *The Barcelona Review*, and *North Dakota Quarterly*. He has a PhD from the Universitat Pompeu Fabra in Barcelona.

Lewis Johnson is an Assistant Professor of the history of art, design and visual culture in the Department of Photography and Video in the Faculty of Communications at Bahçeşehir University. His teaching and research interests include western art from 1750, with specialization in formations of the internationalist avant-garde; the pre-history and early forms of photography; nineteenth century and recent debates about art and technology; word and image practices and theory; pre and post-digital scripto-visual practices; video art; as well as aesthetics and the philosophy of art. His recent publications concern the work of post-war Turkish painters, the gestural and US visual culture (Journal of American Studies in Turkey, 2007); an essay on problematics of identity in post-1990 Turkish art and culture (Identity Theft: The Cultural Colonisation of Contemporary Art, ed. Jonathan Harris, Liverpool University Press, 2008). His research on late nineteenth century Ottoman art, archaeology and visual culture in relation to the politics of ethno-nationalism and the formation of the Turkish republic also continues.

James Lambert is an Australian linguist with over 15 years experience working as a professional lexicographer. He has written and edited dictionaries for the Australian, South-East Asian and International markets and is a world expert on Australian slang. Works include *The Macquarie Australian Slang Dictionary*, 2004, and the Australian entries for *The New Partridge Dictionary of Slang and Unconventional English*, 2005. He currently teaches English as a Second Language in the English Preparatory School at Fatih University whilst doing an M.A. in English Literature and continuing work on his magnum opus: *Across the Zodiac: A Historical Dictionary of Science Fiction.*

Sybille Lammes is Assistant Professor at the Department of Media and Culture Studies, Faculty of Humanities, Utrecht University. She has published on SF film, games and digital cartography. In recent years her main research subjects have been related to the new media and digital culture. Her research programme in computer games examines how games can function as cultural spaces for new spatial and postcolonial practices. Her latest research programme in digital cartography looks at to what extent and how digital maps have altered meanings of media and cartography.

Visam Mansur is a Professor of English literature at Fatih University and Chair of the English Department. He previously held similar teaching and administrative positions in various universities in the Middle East and Europe.

Louis Mazzari is an Assistant Professor in the Department of Western Languages at Bogazici University in Istanbul. He is currently editing a history of Turkish-American relations for Bogazici's Department of History. In 2006, Louisiana State University Press published *Southern Modernist*, Mazzari's biography of Depression-era, civil-rights activist Arthur Raper. Mazzari wrote a new introduction to Raper's *Preface to Peasantry*, in a new edition published in 2005 by the University of South Carolina Press. He was a major contributor to the *Encyclopedia of New England*, published in 2004 by Yale University Press.

Ekaterina Pantcheva is a research associate at the Bulgarian Academy of Sciences. By the time of writing this paper she was teaching English at the University of Architecture in Sofia. She gained an MA in Medieval Studies (CEU, Budapest, 2001) with a thesis on the construction of space in a 14th-century Bulgarian hagiographic text.

Joshua Parker, Assistant Professor of English Language and Literature at Fatih University, has published articles on place in fiction and cultural change in *symplokē*, *Interdisciplinary Studies in Literature and Environment, The International Journal of the Humanities, Urbana* and the *Encyclopedia of African American Literature,* and has contributed a chapter on place in fiction in the forthcoming volume *Hemingway and the Black Renaissance.*

Jerilyn Sambrooke is a PhD candidate at the University of Colorado, and has taught English Literature at Lithuania Christian College in Klaipeda, Lithuania. Her current research concerns the intersection of postcolonial theory with discussions of migration and identity in the context of globalization. Ms. Sambrooke has an M.A. in Literature, Philosophy, and Religion from the University of Sussex and is originally from Alberta, Canada.

Pavel Senderák graduated from an M.A. program at the Department of Politics and European Studies at Palacky University in 2009. His research interests include European integration process (particularly its cultural and language dimensions), and the related identity issues; the changing relationship between national identity, culture, and language; and the position of minorities in Europe (specifically regarding ethnicity, culture and language).

Charlotte Tindale Sexton is currently completing PhD studies at Monash University. She is a writer and artist who published in *Natural Selection, Eyeline, Un Magazine* and *Like, Art Magazine.* She has exhibited at Melbourne galleries including Trocadero, Seventh, West Space and Kings. Her work has been reproduced in *Modern Painters.*

Lucie Tunkrová, graduate of Miami University, Ohio, USA (international relations) and Palacky University in Olomouc, Czech Republic (politics and European studies), PhD obtained in political science from Palacky University in December 2006. She is an assistant profesor at Fatih University, Istanbul, and Palacky University, Olomouc. Her field of specialisation is European integration with special focus on enlargement and theories of the EU's decision making processes. Her other areas of interest are Nordic countries and the process of Europeanization in Turkey. She is a member of a research team working on a 5-year grant titled Czech National Interests in the European Union.

Tania Zuli is a Lecturer at the University of Roma Tre - Italy. Her studies concern Nineteenth-Century Colonial Literature and the Post-Colonial novel. She has published articles on Jane Austen, R. L. Stevenson, Rider Haggard, Arnold Bennett, George Meredith, Anthony Powell and James Joyce. She is author of the monograph *Nadine Gordimer. Strategie narrative di una transizione politica.* She is currently working on a volume on the narrative of Rider Haggard. She is part of the Editorial Team of *Merope*, the journal of the Dipartimento di Scienze Linguistiche e Letterarie of "G. d'Annunzio" University - Pescara.